Emerging Military Technologies

Recent Titles in
Contemporary Military, Strategic, and Security Issues

Emerging Military Technologies

A Guide to the Issues

Wilson W. S. Wong

Contemporary Military, Strategic, and Security Issues

 PRAEGER

AN IMPRINT OF ABC-CLIO, LLC
Santa Barbara, California • Denver, Colorado • Oxford, England

Library of Congress Cataloging-in-Publication Data

Wong, Wilson.
  Emerging military technologies : a guide to the issues / Wilson W.S. Wong.
    p. cm. — (Contemporary military, strategic, and security issues)
  Includes bibliographical references and index.
  ISBN 978-0-313-39613-7 (hbk. : alk. paper) — ISBN 978-0-313-39614-4
(ebook)   1. United States—Armed Forces—Weapons systems.
2. Military research—United States.   3. Military weapons—
Technological innovations—United States.   I. Title.
  UF503.W66 2013
  623.0973—dc23   2012036297

ISBN: 978-0-313-39613-7
EISBN: 978-0-313-39614-4

17 16 15 14 13   1 2 3 4 5

This book is also available on the World Wide Web as an eBook.
Visit www.abc-clio.com for details.

Praeger
An Imprint of ABC-CLIO, LLC

ABC-CLIO, LLC
130 Cremona Drive, P.O. Box 1911
Santa Barbara, California 93116-1911

This book is printed on acid-free paper ∞

Manufactured in the United States of America

Contents

Acronyms

ABL	Airborne Laser
ADS	Active Denial System
AEHF	Advanced Extremely High Frequency
AFRL	Air Force Research Laboratory
ALMHV	Air-Launched Miniature Homing Vehicle
ARPA	Advanced Research Projects Agency (defunct term)
ARPA-E	Advanced Research Projects Agency-Energy
ASAT	Antisatellite
ASBM	Antiship Ballistic Missile
ATL	Advance Tactical Laser
BCI	Brain-Computer Interfaces
BWC	Biological Weapons Convention
C4ISR	Command, Control, Communications, Computers, Intelligence, Surveillance and Reconnaissance
CCW	Convention on Certain Conventional Weapons
CDC	Centers for Disease Control and Prevention
CED	Conducted Energy Device
CIWS	Close-In-Weapons-System
CNT	Carbon Nanotube
COTS	Commercial Off the Shelf
COTS	Commercial Orbital Transportation Services
CWC	Chemical Weapons Convention
DARPA	Defense Advanced Research Projects Agency
DEW	Directed Energy Weapons

DNA	Deoxyribonucleic Acid
DU	Depleted Uranium
EEG	Electroencephalography
EM	Electromagnetic
EMI	Electromagnetic Interference
EMP	Electromagnetic Pulse
EOD	Explosive Ordnance Disposal
ET	Emerging technologies
EW	Electronic Warfare
FAE	Fuel Air Explosives
FEL	Free Electron Laser
fMRI	functional Magnetic Resonance Imaging
FOBS	Fractional Orbit Bombardment System
GM	Genetically Modified
GPS	Global Positioning System
HEL	High-Energy Laser
HMD	Head-Mounted Displays
HPM	High-Power Microwave
HUD	Heads-Up Display
ICBM	Intercontinental Ballistic Missile
IED	Improvised Explosive Devices
IFF	Identification Friend or Foe
INS	Inertial Navigation System
IOC	Initial Operating Capability
JDAM	Joint Direct Attack Munitions
LADAR	Laser Detection and Ranging
LEO	Low Earth Orbit
LIDAR	Light Detection and Ranging
LOAL	Lock-On-After-Launch
LOC	Lab-on-a-Chip
MAD	Mutually Assured Destruction
MANPAD	Man Portable Air Defense
MIRV	Multiple Independent Reentry Vehicle
MRE	Meal, Ready-to-Eat
MTHEL	Mobile Tactical High-Energy Laser
NATO	North Atlantic Treaty Organization

NBC	Nuclear, Biological, and Chemical
NNI	National Nanotechnology Initiative
OPFOR	Opposition Force
ORS	Operational Responsive Space
PAROS	Prevention of an Arms Race in Outer Space
PORC	Performance Optimizing Ration Components
PTSD	Posttraumatic Stress Disorder
RAM	Rocket, Artillery, and Mortars
RF	Radiofrequency
RMA	Revolution in Military Affairs
ROE	Rules of Engagement
RWS	Remote Weapon Station
SAM	Surface-to-Air Missiles
SBL	Space-Based Laser
SCADA	Supervisory Control and Data Acquisition
SDI	Strategic Defense Imitative
SFW	Sensor Fused Weapon
SHORAD	Short-Range Air Defense
SLBMs	Submarine-Launched Ballistic Missiles
SSTO	Single-Stage-to-Orbit
TBI	Traumatic Brain Injury
TDNDS	Transdermal Nutrient Delivery System
THEL	Tactical High-Energy Laser
TSTO	Two-State-to-Orbit
UAS	Unmanned Air System
UAV	Unmanned Aerial Vehicle
UCAS	Unmanned Combat Air System
UGS	Unmanned Ground System
UGV	Unmanned Ground Vehicle
USV	Unmanned Surface Vehicle
UUV	Unmanned Undersea Vehicle
WMD	Weapons of Mass Destruction

CHAPTER 1

Introduction

War and military operations other than war will endure as viable policy options, though these are fraught with risks and costs. Technology has the potential only to modify these realities, but not to eliminate them completely. As the state of the art in technology changes, it is always important to consider the effects of new technologies on military options. Today's emerging technologies (ETs), such as advance computing, nanotechnology, and biotechnology, are poised to change the world and therefore the shape of international relations as well as military affairs.

Technological change in many areas is quickening. Exponentially growing computer power, government support for science in the United States and internationally, and the relatively new factor of consumer demand for high-technology goods are among the factors accelerating many areas of technological development. Beyond improving familiar and comfortable goods and services, technological development can bring into being seemingly new and different technologies, ETs. In general, an ET is one that has the potential to be disruptive to society by changing the existing norms. Those that can successfully adapt to the norms that emerge from the aftermath will be poised to be the winners in the new age. ETs therefore represent opportunity for some to make rapid advancements in the world, while for those comfortable in their current status, it is an opportunity to fall. The opportunity for rapid changes in the comparative status applies to individuals, organizations, businesses, and nations. For individuals and companies, this has the potential to bring fortune or ruin. However, in the anarchical international system, changes in norms and comparative power are often accompanied by conflict up to the level of systemic (global) war.

The present technological era is alternatively known as the digital, computer, or information age. The foundations of this era can be traced to various limited-function calculation devices of antiquity, but more commonly these are linked to the early programmable computers of World War II. Its impact on the general society, as well as on warfare, accelerated in the 1970s and 1980s with the rise of the semiconductor industry. Displacing fragile and expensive vacuum tubes, semiconductors allowed for not only reliable computing but also low-cost computing. This proliferation of computer power, in turn, spawned the digital age. At present, the world is still adapting to the changes brought on by the information technology revolution. Governments, industries, and individuals are still in the midst of tackling the many policy questions and societal challenges brought on by the seemingly overnight emergence of the information age. Policy issues such as digital privacy, cyber bullying, and digital rights management were not salient issues in the early 1990s. As there were societal winners and losers in the transition from an industrial society to an information society, the next ETs will have a similar capacity to change the rules of the world.

Advances in semiconductors and other ETs needed to bring on the information revolution would also make possible the reliable and relatively low-cost precision-guided munitions (PGM) of the last Revolution in Military Affairs (RMA). ET, in a defense context, appeared in an episode of a British satire, *Yes Prime Minister,* as a prominent option for defending the realm in the 1980s, though the hapless fictional prime minister, Hacker, did initially confuse ET with the sci-fi movie. At the time, precision weapons were somewhat unproven to military authorities, despite some success in the Vietnam War, and probably did sound like science fiction to policy makers unconvinced of their potential. Since the 1980s, precision weapons have become a major part of the conventional warfare paradigm, being proven to military officials, policy makers, and the general public in conflicts such as the 1991 Gulf War, as well as the Kosovo conflict at the end of the last century.

Military ET, having a limited customer base, must compete with existing concepts and systems for limited funding. Not only must new technologies be procured and maintained, but they must also be developed. The United States is the preeminent military power, but this lead has come as a result of a large expenditure on research and development. Internationally many of the technologies associated with U.S. military power have been replicated, often at much lower costs, because the United States has already borne the expenses of proving the viability of these technologies. Both Russia and China are developing their own stealth aircraft, but neither had to go through the trials of proving the utility of the concept.

In addition to the challenges to the preeminent power within the bounds of the dominant military paradigm, there are also those that will seek to find specific counters to the tactics and technology that represent the current state of the art. Stealth aircraft are an expensive technology and are possibly more expensive than the countermeasures including counter-stealth measures such as more sensitive radars coupled with improved analysis software and guided surface-to-air missiles (SAMs), which are already generally cheaper than manned combat aircraft. Ironically computers allowed the calculations needed to produce radar-defeating shapes (that could fly) and are now poised to allow lower-cost defenses to challenge such invisibility. Insurgency tactics and terrorism, concepts central to today's definition of asymmetrical warfare, also by definition seek to indirectly challenge conventional military power.

Intentionally pushing technological change itself can jeopardize any comfortable advantages in the military power a nation may possess. In the debate over the construction of H.M.S. *Dreadnought*, the world's first all-big-gun battleship among other innovations, it was feared that its appearance would make the existing superiority in pre-dreadnought battleships held by the Royal Navy obsolete. On the other hand, the basic technologies and operational knowledge that led to this new form of capital ship had proliferated; *Dreadnought* has contemporaries under consideration in the United States and Japan.[1] In the ongoing debate over the deployment of antisatellite (ASAT) weapons, there is the perspective that overt weaponization of space would jeopardize current U.S. superiority in space-force-enhancement capabilities by inspiring others to weaponize space as well. A competing perspective is that the space capabilities that the present-day U.S. military might is built on are only natural targets, the recognition of which would invite an asymmetrical response[2] through space weapons. Throughout history, seeming patterns of military innovation and imitation have been upset by asymmetrical response and outright technological surprise.

Dual-Use and Spin-Ons

Dual-use refers to technologies and goods that may be used both by militaries and civilian industries. These include mundane items such as diesel engines and computer chips. Specialized technologies such as those found in the civilian chemical and pharmaceutical industries have militarized applications such as chemical and biological weapons production—a fact that has made limiting the proliferation of weapons of mass destruction (WMD) very difficult. The ongoing alarm over Iran's, allegedly peaceful, nuclear-enrichment activities is a reminder that nuclear technology is dual-use as well, despite the greater attention and fear this technology holds. Many of today's ETs are dual-use in nature, meaning that these too may proliferate under the guise of "peaceful use," despite their effect on the military balance of power.

Spin-offs of military technology are well known, with examples such as jet engines, computers, and duct tape on the list of dual-use products with military origins. Spin-on military technologies are not quite so well known. The first trains, airplanes, and automobiles were civilian machines first and only later found strategic and tactical uses. Often new technologies gain acceptance in the civilian world first due to the generally more-benign operating environments found off the battlefield. The world's first tanks, another spin-on military technology, certainly had their share of mechanical difficulties and limitations, only made worse by having to be discovered and debugged in the misery of World War I. In military use, the horse was able to retain an edge over mechanization for a few years after World War I—it did have the advantage of millennia of operational experience over the relatively recent invention of the internal combustion engine and the caterpillar track.

Many of today's latest technological developments are often found in the civilian world with no real equivalents in the military. Deployment of personal digital communication devices in the military still lags behind the near-ubiquitous adoption of cell phones in the civilian world—military requirements and acquisition processes have led to a different marketplace for military digital radios. That said, there is interest in using off-the-shelf smart phones for some applications in the military.[3] With the ETs, there is a race between the military and the medical communities to field the world's first production exoskeleton.

Technology and War

Since at least the shock of *Sputnik* in 1957, the measure of a nation's power has included its capacity for science. Technology is, however, more than science; it also encompasses the ability to produce something useful—an ability to convert abstract scientific discovery into something tangible. These two capabilities have intertwined throughout history to produce national wealth and security. Even with many high-technology industries being transnational, a nation's capacity to support a high-technology economy remains such a measure.

Definitions for technology include tools and the practical use of tools. Weapons are also tools, and their practical employment is codified in doctrines, tactics, and strategies. While it is often useful to have new weapons filling existing roles, full exploitation of a new military technology requires the development of unique and specific doctrines suited for the opportunities created. The early tanks of World War I were of questionable value[4] without the effective tactics developed later in the interwar period. The correct usage of tanks did not become clear until World War II, during which the German blitzkrieg doctrine proved superior to those employed by France and Britain in 1940. The Allies and Soviets later employed many of the same tank doctrines to push Hitler's forces back into Germany.

New technology does not always lead to correct policies, as some technologies can be dead ends. Toward the later part of World War II, German tanks, such as the Tiger, were more advanced as individual pieces of technology, though this in itself hamstrung the war effort by being too advanced to be easily produced and maintained, arguably contributing greatly to Germany's defeat. By this time, the industrial capacity of the United States and Soviet Union were producing large numbers of simpler tanks, such as the M4 Sherman and T-34, which simply overwhelmed the small numbers of German super-tanks.

World War II represents the beginning of the conscious and deliberate link between technology and military power. The scientist came to prominence with the employment of revolutionary technologies such as radar in warfare, something that the Royal Air Force (RAF) had been able to exploit during the Battle of Britain to coordinate its limited number of fighter aircraft to blunt the previously unstoppable Luftwaffe. On the other side, the Nazi leadership became enamored with the creation of Wunderwaffe, wonder weapons, to stave off defeat toward the end of the war. These included jet aircraft, advance submarines, and guided missiles, all deployed in various stages of technological readiness (or lack of it).

That is not to say that a connection between technology and war did not exist before World War II. In fiction, the original Sherlock Holmes story, "The Adventure of the Bruce-Partington Plans," involved the theft of plans for an advanced-technology submarine. The submarine was an ET in the late 19th and early 20th centuries, with doctrines still under development and fearful naval experts as well as pacifist elements of civil society calling for its prohibition. The long stalemate of World War I, characterized by the doctrine of trench warfare, was itself the product of the industrial revolution products such as the machine gun and barbed wire. Historians have reached back further to compare dissimilar weapons and tactics employed by groups in conflict, perhaps as a reflection of the current preoccupation with comparing the capabilities of today's military forces. However, in these past conflicts, technology and scientists were not celebrated (or demonized) to the extent they were in the aftermath of World War II and in the present day.

In the United States, there was the Manhattan Project[5] to raise the profile of the scientist in war. Many Manhattan Project scientists became prominent, such as Edward Teller, who later became a vocal supporter of the Strategic Defense Initiative (SDI). Others, such as Leó Szilárd, became prominent in the disarmament movement. The initiation of the Manhattan Project itself was pushed along by the already prominent scientist Albert Einstein: the

Einstein–Szilárd letter warning of the potential for a Nazi-developed atomic weapon was instrumental in convincing the United States to undertake atomic weapons development.

In the prelude to the Cold War, the dismantlement of Nazi Germany, Allied and Soviet forces raced to collect German scientists and examples of military technology. Many of the weapons of the early Cold War were based on this captured body of work and expertise. Both the Mig-15 and F-86 Saber owe their wing design to the German research. In the United States, former Nazi rocket scientist Wernher von Braun gained celebrity status promoting the U.S. Space Program, an offshoot of his work for the U.S. Army.

The Soviet Union had also infiltrated spies into the Manhattan Project, greatly accelerating its own efforts to develop the bomb. The Cold War espionage war between the East and the West highlighted the importance of military technology as both sides attempted to collect as much information as possible on each other's latest weapons and platforms. U.S. missile tests were monitored by Soviet surveillance vessels sitting off the coast in international waters. Spy and counter-spy technology pushed the limits of technology, with miniature cameras, passive electronic devices, and code-breaking supercomputers. Descendants of these secret devices are today's web cameras, radio frequency ID (RFID) tags, and home computers.

A defining moment of the Cold War technological race was the launch of *Sputnik*, the world's first artificial satellite, in 1957. Unlike the Soviet Union's first atomic bomb test, which could be attributed to espionage, the launch of *Sputnik* did, for a while, give the impression that the United States' technological prowess had fallen behind—"There was a sudden crisis of confidence in American technology, values, politics, and the military."[6] This event was an ET "Pearl Harbor" due to the general surprise, shock, and later mobilization of the United States that resulted. The response included massive investment by the United States not only in the space program, but also in many areas of fundamental scientific research, as well as in education. Education remains a national security concern, as without scientists and technicians, the nation would fall behind in other technology areas, including those critical to protecting the country. At great cost, both financially and in the lives risked, the United States won the "space race" with the first manned landing on the Moon in 1969.

This climate of seemingly desperate technological competition sparked by *Sputnik* was the impetus for the creation of the Advanced Research Projects Agency (ARPA), later Defense Advanced Research Projects Agency (DARPA). DARPA has the mandate to:

> . . . maintain the technological superiority of the U.S. military and prevent technological surprise from harming our national security by

sponsoring revolutionary, high-payoff research bridging the gap between fundamental discoveries and their military use.[7]

While the civilian spaceflight program has languished[8] since the heydays of the space race, DARPA continues to foster technological development by funding breakthrough, often seemingly unbelievable, research and development programs. The workings of the Internet, an ARPA invention, remain a mystery to the majority of people outside computer tech support. All of the ETs discussed here have publically known interest from DARPA.

The dominant military technology of the Cold War was nuclear weapon, with both sides eventually acquiring tens of thousands of warheads. These warheads armed what seems by today's standards to have been several rapidly replaced generations of bombers, missiles, and even artillery. Military aeronautics in the early Cold War went through a period of rapid development, when procurement programs would buy small quantities of a system with the expectation, and usually the reality, that something better would be available in a few years. Service lengths of less than a decade were not uncommon as systems rapidly became obsolete in the face of new technology. Today the development programs of aircrafts such as the F-22 and F-35 are decades in length, with the latter still not in service despite being officially started in the early 1990s with the merger of several similar programs that were already in progress.

There were, however, limits to this trend; technology and finances often present barriers to further performance increases. The U.S. B-52 is one example, as is the long drawn out history of attempting to find a successor to the M16/M4 rifle. At present, the United States is again considering a replacement bomber program for the B-52, an aircraft that first flew in 1954 and presently is expected to be in service beyond the year 2040.[9] Consumer computer technology has arguably already run into something similar, with clock speeds no longer jumping as rapidly as they were only 10 years earlier due to physical limitations, such as heat dissipation, coupled with the average computer user not actually needing much more for applications such as e-mail, web browsing, and word processing.

During the Cold War, there emerged a very public debate over the usability of weapons. Outside the disarmament movement, there was a degree of uncertainty among actual defense academics over the utility of nuclear arms. At the time, the Soviet Union possessed a numerical advantage in armored forces, and the (conventional) warfare paradigm that emerged out of World War II involved numbers. The reality of Soviet armor superiority would have put the West in a precarious strategic position, if not for its nuclear arsenal, including the controversial tactical nuclear weapon concept. Among the problems with tactical nuclear weapons was the unclear line between them and strategic weapons. It was feared that tactical nuclear weapon employment to stop the clichéd Soviet armor invasion of West Germany through the Fulda Gap would escalate to an all-out nuclear war. Linked to this was the question of whether the United States would risk nuclear attack on itself to protect Western Europe. Additionally, there was the problem that defensive use of nuclear weapons would mean using them on the territory being defended. The sheer power of nuclear weapons led them to being something of an "unusable" weapon, too powerful for the majority of post–World War II conflicts, but still needed to prevent others who did not share that view from using them with impunity.

While the United States and the North Atlantic Treaty Organization (NATO) were debating the merits of the tactical nuclear weapons doctrine, and nuclear deterrence was becoming the core strategic reality governing East–West relations, the computer revolution was taking place, allowing the creation of unquestionably effective PGMs. The force-enhancement effect of precision-guided weapons allowed the United States and NATO to provide an additional counter to Soviet's numerical superiority. While PGM technology was maturing, the West retained the option to go nuclear, and ultimately nuclear weapons remained the core military concern during this transition period. Presently, nuclear weapons are of diminished but still important status,[10] and hi-tech conventional warfare is an entirely usable option as has been proven since the 1990s.

An important consideration in the choice to use military forces is risk, something that technology can modify but cannot remove entirely. For some, the risks of nuclear technology, specifically mutually assured destruction (MAD), were a factor in superpower stability, which prevented major systemic war from breaking out during the Cold War. There were concerns raised on both sides of the Iron Curtain over the potential for new technologies such as improved missiles or anti missile systems to be destabilizing—to change the risk calculus of nuclear war, thereby making the nuclear option more likely. For others, the risk of global nuclear annihilation was generally unacceptable, leading to calls for disarmament.

Among the force enhancements conferred by precision-guided weapons is survivability of the attacking force—effective reduction in risk for individual soldiers arguably increases the tolerance of a society toward warfare due to diminished risk. Precision weapons increase survivability by reducing the number of sorties to achieve the same effects. Standoff precision weapons, such as drone aircraft, increase survivability by allowing a greater distance between the target and the attacker. However, society's tolerance, or intolerance, to risk now extends to that of unnecessary application of force—risk aversion toward collateral damage.

To be clear, precision warfare was developed primarily as a force enhancer—to allow relatively small forces to accomplish tasks that required in past much larger investments in force. The humanitarian aspect only came to the fore later as these weapons were used in conflicts other than total war, and indeed in military operations other than officially declared war. Arguably, increased precision as a simple force-enhancement capability for use against conventional military foes has diminishing returns because hard military targets do require a minimum of force to destroy, though the precision available would, on the surface, appear to be sufficient. However, hard military targets are not the only threats, and the United States and its allies have continued to invest in ever smaller and more precise weapons to counter smaller threats, possibly with the next wave of ETs down to individuals in a crowd.

That is not to say that these efforts are always successful; like all real-world implementations of concepts, there must be room for accidents, unforeseen events, and mistakes, such as misidentification of targets. Indeed the lower destructive power of today's focused weapons has led to charges that their very precision has made their use all too alluring for politicians—warfare made acceptable due to overhype that war can be fought without collateral damage. The argument is that, unlike nuclear weapons, precision weapons are all too "usable," though whether this is a good thing depends on one's personal view on war as an instrument of policy. However, in a reality where not all threats and international crisis can be dealt with by discussion and negotiation, the military options will remain viable. Whether by design, accidental by-product, or requirement forced by a media-saturated society, it is likely that there will be continued investment in increasingly precise weapons. This, in turn, can only benefit civilians caught in the cross fire. While not all collateral damage can be prevented, at the very least the paradigms of the current and near-future military affairs will continue to emphasize greater discrimination between friend, foe, and bystander and therefore provide more options toward reducing such damage.

Today's Military Emerging Technologies

In the second decade of the 21st century, several emerging technological areas expected to reshape civilization have been identified. Three of these technological fields are advance computing, nanotechnology, and biotechnology. All three of these are in general dual-use, meaning that the only certainty with these technologies is that there will be military applications. However, the history of technological prediction, futurism, is littered both with failed dreams and unexpected challenges that have slowed development. The long and troubled development of ubiquitous (military) space access and directed energy weapons (DEWs) round out any discussion on emerging military technologies.

Chapter 2: Ubiquitous (Military) Space Access

Prior to the current information age was the space age, another era of technological optimism. The harsh financial and related technical challenges of space access have ended most of those dreams, though for the U.S. military this limited access has proven sufficient to build the foundations for contemporary U.S. military dominance. Exotic propulsion technologies have offered to change the price of spaceflight, keeping many space age dreams alive across generations of enthusiasts. In the near term, however, streamlined operations, advance materials, and control technologies offer to slash costs via old-fashioned rocketry. If either form of ubiquitous space access becomes viable, there will be security implications. Enhanced space access does not exist purely for its own purposes, and unlike many space dreamers, the military has been funding space access research since the very beginning. Although results for many have been too slow, many of the technologies being invested in today are finally reaching maturity. Although space access is expected to remain expensive, any lowering of the cost will affect the military first. Despite the present-day constrained funding environment, space remains the "highest of the high ground."

Chapter 3: Directed Energy Weapons

Probably the most unabashedly military of ETs, DEWs have also been stuck in emerging status for decades. Demonstrated under laboratory conditions, the deployment of DEW systems have been stalled by many operational challenges. Convergence with other technologies may, however, produce a robust enough system worthy of being called a weapon. As an ET, the form of DEW systems and doctrines remains to be seen. Will it be a niche capability useful in only a few specialized roles, or will it come to be commonplace as science fiction often predicts? There is also the reality that physical forms of delivering destructive energy will be difficult to displace from the battlefield, possibly leaving DEW again as nothing more than an overhyped light show.

Chapter 4: Advance Computing

As computer technology advances, weapon systems have the potential to gain greater autonomy. Depending on what is included, robot weapons are either an ET, or an old concept. While robotics does have much to offer, there is also opposition to certain aspects of the weapon-bearing robot concept. This goes beyond the many technical problems of entrusting an autonomous machine with a lethal capacity and will discuss political, legal, and possibly ethical roadblocks. The existence of a potentially "disposable" soldier in the form of a robot or remotely operated weapon has both opportunities as well as controversy.

Chapter 5: Nanotech

Nanotechnology, the application of engineering on the nanometer (one billionth of a meter) scale is often touted as the next "big thing" for society at large and already has a host of emerging, short-term, and long-term military applications. Putting aside the hype, nanotechnology can be regarded as the basic sciences and engineering simply done at the finest levels of precision—many of the products of this field are simply refinements on existing goods and services. In the near term, nanotech represents a new form of material science, which in the 21st century must face more mundane scrutiny over its potential unintended side effects, such as environmental impact.

Like many of the ETs, there is international competition in this field, including in military applications. The forms of nanotechnology currently pursued are largely low-energy processes that may be easily hidden. Futurists even envision the proliferation of nanotechnology tools to empower those who previously had none. However, not everyone shares this vision of a bright shiny future assembled by nanotechnology, and the United States may be surprised by who will be able to use this technology for nonpeaceful applications. While both nuclear and nanotechnologies are in general dual-use, nanotechnology has not acquired the same stigma, though there are those who argue that nanotech represents as great, or even greater, an existential (extinction) threat to humanity than nuclear proliferation. These worries range from nanotech contributing to humanity simply being replaced as the top species on earth, to the now largely defunct "grey goo" scenario. The ease at which nanotech is expected to spread, essentially for economic and perhaps even charitable reasons, in a security and defense mindset represents a wide variety of potential new threats or spins on old ones.

Chapter 6: Biotechnology

The emerging biotechnologies promise minute control over the functions of life and death. In between these extremes are opportunities to enhance the human condition, and to enhance the individual soldier. Improved military health care will only be a starting point for the things that the convergence of biology, computers, and nanoscale manipulation will allow.

Biotechnology opportunities will, however, raise many troubling questions about where this fine control over life itself will lead. The military will have different answers to the many societal questions raised by advanced biotechnology. In an all-volunteer force, would some applications such as genetic screening be more acceptable than others, say for an insurance provider? Military service involves sacrifice in the defense of one's country, but will the opportunities provided by biotechnology be asking for too much on Western society? This is a proliferating technology, so competitive and security challenges posed by other nations' exploitation of biotechnology may take this decision out of the hands of the general populace.

Convergence

The seemingly unpredictable nature of technological development is reflected by perhaps the ultimate expectation of the continued exponential growth in computer power combined with the other ETs, the "technological singularity." Although sci-fi sounding, a singularity denotes points where the known laws of the field, such as physics, fail, a definition that in turn seems to tinge this real-world concept with fantasy. Up until the singularity of a black hole, the point where so much mass has collected that escape velocity exceeds the speed of light, physics can explain it. At and beyond this singularity, things become open to much more speculation. The many possibilities of the technological singularity have spawned its own subculture or subset of posthumanism philosophy, "transhumanism," where superior computer intelligence brought on by the convergence of many ETs is used to either augment or replace ordinary humans, often with all sorts of utopian outcomes for the human condition. Those who believe that this event will occur—and there is a lot of debate on how much computer power is needed, and even if computers are suitable for replicating the mind—expect an unimaginable amount of societal and technological change once humanity can produce something more capable than natural-born human intelligence. Alternatively, much of today's dystopian science fiction concerns a future where this and other ETs have run amok, and the rise of artificial intelligence is not beneficial to humanity. Clearly not everyone agrees that intentionally working toward creating humanity's replacement is a wise idea.

Although the notions encompassed by the concept of the technology singularity are largely beyond the time frame under consideration here, not everyone would agree. A range of technologies are converging that give hope to those who want to soon see the miracles of a postsingularity world. Computer capabilities, as given by number of transistors on an ordinary computer chip, are increasing at an exponential rate with no end expected in the near term. Research into advanced computer capabilities, such as an ability to learn and reason, is being fuelled by both business and international competition. Already software mimicking components of living brains, neural nets, are in use to solve a range of computational tasks that other techniques cannot handle efficiently. Among the things neural nets are used for is machine learning, the capacity for a computer to be trained and gain experience. For both defense and civilian applications, the ability to learn is a possible route to producing autonomous machines capable of operating outside controlled environments.

The study of the human mind, aided by the rapid and ongoing development of computerized tools, is increasing our understanding of all aspects of intelligence. Successful technology investor Ray Kurzweil has predicted that by 2030, still within the near to midterm, computer technology and biotech would have advanced to the point where it would be possible to scan and then replicate a functioning human brain as a computer simulation. A 2007 U.S. Congress study put the necessary technology for singularity sometime after 2020. In the context of U.S. government's activities, there are, as mentioned earlier, military procurement programs that have taken longer to reach initial operating capability (IOC) than the estimated minimum of 13 years for the technology base needed for the singularity to appear. Quite clearly much is expected from the Emerging Military Technologies.

Notes

1. Robert K. Massie, *Dreadnought: Britain, Germany, and the Coming of the Great War* (New York: Ballantine Books, 1991), 487.

2. The symmetrical response to U.S. space power is the replication of space-force-enhancement capabilities. This in turn would build a case for the United States to weaponize space, as it would no longer have an advantage, and therefore, there would be utility in being able to disrupt the use of space by others.

3. United States Army, "Connecting Soldiers to Digital Applications," *Stand-To!*, July 15, 2010, http://www.army.mil/standto/archive/2010/07/15/.

4. Keir A. Lieber, *War and the Engineers* (Ithaca: Cornell University Press, 2005), 101–3.

5. It must be noted that the Manhattan Project had British and Canadian involvement.

6. Paul Dickson, *Sputnik: The Shock of the Century* (New York: Berkley Books, 2007 edition), 4.

7. Defense Advanced Research Product Agency, "About," http://www.darpa.mil/About.aspx.

8. Notwithstanding the remote but real threat of impact by natural space debris, space exploration simply does not have the same priority as the practical needs of defending the state. This, however, does not stop spaceflight enthusiasts from bemoaning the large disparity between the Department of Defense's budget versus that of NASA. That said, it must also be remembered that NASA's mandate goes

beyond spaceflight and includes research into flight phenomena wholly within the atmosphere.

9. United States Air Force, "Factsheets: B-52 STRATOFORTRESS," http://www.af.mil/information/factsheets/factsheet.asp?id=83.

10. The United States, United Kingdom, and France maintain a policy of ambiguity over whether they would be first to use nuclear weapons, though recent policy changes by the Obama administration, such as the April 2010 *Nuclear Posture Review*, have indicated that for nonnuclear nations, with a few exceptions, the nuclear attack option was not a policy option. This document, however, does not rule out first use of nuclear weapon given the appropriate adversary and circumstances. The People's Republic of China (PRC) and India have No First Use policies toward nuclear weapons; however, with closed societies, such as the PRC, there is some uncertainty over the credibility of this policy.

CHAPTER 2

Ubiquitous Space Access

It is fair to say that space activities, military and civil, in the early 21st century have not turned out as imagined by futurists at the dawn of the Space Age. Although it is true that many high-profile concepts from the early Space Age seemed stalled, other applications have become the unseen, and out of mind, background of modern society and of modern warfare. A large factor driving space use to the forms seen today has been the economics of launching payloads into low earth orbit (LEO). The high cost of spaceflight has given a strategic flavor to most space applications. The development of ubiquitous space-access technologies, the capability for low cost, and robust and reliable orbital launch, have the potential to disrupt the shape of military space power. However, the technical challenges of space access are great, and have kept many promising technologies stuck always "a few years in the future." Despite a long list of failed dreams, ubiquitous space access endures as a desired emerging military technology.

Space launch has a proven technology base. The problem is that thus far it is an expensive technology, which while not exactly experimental, is far from casual. These realities of spaceflight have until recently meant that space programs are strategic in nature. The euphemism for spy satellites, "national technical means," alludes to the strategic nature of these platforms. Individually, many of these platforms are worth billions of dollars. Prominent programs making their way to orbit today, such as the MILSTAR 3/Advanced Extremely High Frequency (AEHF) communication satellite constellation have been in the making for years. In the case of the AEHF program, initial contracts for program definition were awarded in 1999,[1] actual development in 2001,[2] and the first launch of at least three satellites only occurred in August 2010.[3] In addition to the United States waiting for

the expanded secure communications provided by the AEHF are its partners in this satellite program: Canada, the Netherlands, and the United
Kingdom.

Like other similarly priced national assets, the high cost of military space
infrastructure is balanced against the capabilities this infrastructure provides.
During the Cold War, space-based strategic reconnaissance and missile early
warning were part of the information system that allowed nuclear deterrence
to be a viable strategy. Today U.S. military space power, although very costly
in itself, is providing savings through force-enhancement services to forces
within the atmosphere. The mass use of the NavStar global positioning system (GPS) and satellite communications are the foundations of quick and
agile U.S. global power.

Ubiquitous space access, or space access that is robust, reactive, and reasonably low cost, would potentially expand the number of missions that may
be undertaken, such that lower levels of the security and defense hierarchy
(theatre commanders for instance) could have operational control over space
assets. This then leads to the subject of operational responsive space (ORS).
What constitutes ORS, like the concept of military transformation, encompasses different things to different groups. Among the more common characteristics for ORS, are efforts to lower the cost of space missions and to shorten
the timescales involved. Interest in this capability has resulted in the formation by the U.S. Department of Defense of a Joint Operationally Responsive
Space Office at Kirkland air force base in 2007.[4] Some degree of ubiquitous
space access, for the military at least, would be an enabling capability for
meeting ORS-type objectives.

A Brief Bit of Rocket Science

The vertical component of military operations above sea level can be con-
ceptualized as being made up of three layers: the lower layer (where conventional airpower exists), outer space, and an in-between, "near-space," layer.[5]
These are not the divisions used by atmospheric science,[6] which define layers
based on specific physical characteristics. Instead the three layers to consider
make up a greatly simplified model for demarcating military operations above
the earth's surface, based largely on the state of available aerospace technology today.

The lowest level of this simplified model is where conventional airpower
can exist. At these altitudes, the atmosphere is thick enough to: (1) easily generate aerodynamic lift, (2) support the operation of common aviation engines,
and (3) limit achievable velocities. In the brief history of heavier-than-air
flight, what has been "conventional airpower" has expanded with technological change. The gas turbine jet engine, an emerging military technology of

the 1930s and 1940s, expanded the envelope for airpower from the limits of piston engines and propellers to the current state of aviation. The breaking of the sound barrier is another past demarcation for "conventional airpower." Presently the operational envelope of military aviation is limited to a speed of approximately Mach 2.5, or two and a half times the speed of sound. The now out of service Blackbird family of aircraft represents the extremes of this envelope with 2,000 miles per hour plus maximum speed, and an 85,000-feet-plus service ceiling.[7]

A principal characteristic of the next layer of the military use of the vertical realm, the near-space region of the atmosphere, is the difficulty in sustaining flight in this region. As altitude increases, atmospheric density decreases, resulting in less lift and less oxygen for gas turbine engine operation. Air resistance and the accompanying heat generated by hypersonic flight, while also decreasing with altitude, remain problems until significantly higher altitudes are reached. Launch vehicles, which thus far have been powered by self-contained rockets, fly trajectories meant to quickly pass through this layer into the emptiness of outer space.

The lack of activity in near space may change with aviation technology. Balloon technology is being proposed as a low-cost method to access these very high altitudes, though without any fine control (due to the limited steering options of ballooning). Moving up the scales of cost and technological challenge are very high-altitude airship and very large wingspan unmanned air vehicle (UAV) concepts. Both lighter-than-air flight and the proposed large UAVs are incapable of high-speed flight, their sedate airspeeds in turn being the selling point of persistence over an area of operation.

Outer space, or the military region above the earth's surface where orbiting is possible, is the final layer in this simplified model. There is no internationally recognized standard for where airspace ends and space begins. International treaties, such as the *1967 Outer Space Treaty* (OST), do not define where outer space begins. In the United States the criteria for spaceflight have been dependent on the context and/or the agency involved. At present in the United States, to earn astronaut wings, one must achieve an altitude of at least 50 miles (80 kilometers).[8] Another common notational boundary for space is the Kármán line—100 kilometers altitude. Theodore von Kármán calculated that at around 100 kilometers of altitude, the speed needed to generate enough aerodynamic lift to support an object was higher than the velocity of an object in orbit at that altitude. As a result, the object at orbital velocity would be able to avoid hitting the earth without the need for lift generated by aerodynamics. The uneven nature of the atmosphere and the minutia of aerodynamic flight have led to some variations wherein orbital mechanics supersedes aerodynamics. Also, while lift may be negligible at 100 kilometers of altitude, air resistance would prevent long-term unpowered orbiting. Nonetheless, 100 kilometers is close enough to calculated values and is easy enough to remember.

Operationally, the lowest of stable orbits are several dozen kilometers above the Kármán line, meaning LEO is a handy benchmark for ubiquitous space access. Successful orbit is dependent on achieving a high enough velocity perpendicular to a line from the spacecraft to the center of the earth such that while the object is being pulled back toward earth, it falls in such a way that avoids hitting the earth, resulting in a circular path. At extreme altitudes the very thin atmosphere, relative to sea level, permits these velocities to be achieved and sustained without propulsion. Lack of atmosphere density also means a lack of the high temperatures and other destructive forces associated with high speeds within the lower atmosphere. In the domain of conventional airpower, Mach 3 is still remarkable to this day due to the difficulty in achieving this performance. When orbital velocities are expressed in the somewhat problematic unit of Mach numbers,[9] Mach 25 is the usual figure for the minimum needed to achieve orbit. From LEO, a host of low-power maneuvering options, such as ion and solar sails, are under investigation for long-duration orbital missions. Dropping back into the upper reaches of the atmosphere to maneuver is something of an emerging military technology in itself—hypersonic gliding has been proposed for decades as a means to both extend range and to allow for large changes in flight path relative to the locked nature of orbital mechanics.

In general, orbital altitude and orbital velocity is achieved via rocket technology. Rockets are an implementation of Newton's third law: for every action, there is an opposite and equal reaction. In the case of a rocket, propulsive thrust in one direction is the reaction to a mass ejected in the opposite direction. The specific propulsive force is a product of the mass and velocity of the expelled reaction mass. A pure rocket is a self-contained system, carrying onboard all the propellant needed for flight. A rocket can be as simple as letting pressure force out a working fluid, or as complex as using electrical and magnetic fields to propel subatomic particles. For ground to LEO access, rockets burning fuel and oxidizer from onboard supplies have been dominant.

A daunting amount of propellant, fuel and oxidizer, is needed to propel a payload from the notional standstill of the launch pad to orbital altitude and velocity. Konstantin Tsiolkovsky's rocket equation relates a vehicle's capacity to change velocity, its delta-V, to its engine performance and its change in mass after a velocity change. A mass change of 90 percent is needed for a single-stage-to-orbit (SSTO) launch vehicle based on today's chemical rockets. This mass change, or mass fraction, represents the mass of propellant that must be expended to achieve orbit, and the remaining 10 percent, or less, represents the engine, propellant tanks, structure, flight control, and finally payload. Although this is certainly not a welcome number from an engineering or at first glance an economic standpoint, the SSTO concept does not cross over into the impossible. Barring alternatives to rocket-powered SSTO technology, orbital access is a matter of safely burning enough fuel and oxidizer. Advances that produce lighter vehicles and engines, material science in particular, may even make this feasible in the near future.[10]

Not wanting to wait for incredibly lightweight structures and engines, the real world of orbital access has depended on multiple stages, both stacking them on top of one another, as was done on the *Saturn V,* or through the use of parallel stages that may burn for a portion of flight together, as is done with the space shuttle. The staging concept is attributed again to Tsiolkovsky—the multistage variant of his rocket equation produces significantly higher nonpropellant mass fractions. Staging breaks the problem of reaching orbital velocity into parts. After counteracting gravity and drag, each stage is responsible for only part of the velocity change needed to get up to orbital velocity. Each stage only has to accelerate the payload, the stage itself, its own propellant, and any later stages with their propellant. Additionally, each stage can be tailored for the conditions found in the portion of the trajectory in which it is active. There are differences in the specifics of rockets meant for use in the lower atmosphere and those used in the near vacuum of LEO. The use of multiple stages allows significant increases to the nonpropellant mass fraction, but implies greater cost due to essentially needing multiple vehicles, each of which can unexpectedly end a mission through individual failure. A pure rocket SSTO is feasible today, but would be very large when compared to a multistage launch vehicle of the same payload lifting capability, possibly to the point where to safely fly, it would be more expensive than simply assembling a stack or cluster of smaller less dangerous stages.

Rocket science draws from all other disciplines. Broadly speaking, this includes a fair amount of political science as the history of the Cold War and the bibliographies of leading spaceflight luminaries Wernher von Braun and Sergey Korolyov (Sergei Korolëv) demonstrate. Cold War politics, and the funding it enabled, was key to early Space Age progress. Although there is great potential for a near peer, usually China, to invoke thoughts of another "space race," the budgets for space activities is presently much more restrictive. These shrinking budgets have brought on calls to develop cheaper space access, and cheaper space missions. The basic rocket-powered multistage launch vehicle dates from the 1950s, and a host of technologies seem to be only a few years away from being ready to not only supplant it but also spark a revolution in space access by lowering cost—however, only after expensive R&D efforts. The other means to lower space-launch costs in the near term would be smaller incremental improvements to familiar multistage rocketry. This then leads to a somewhat political battle between those who favor high payoff, but high-risk technologies such as air-breathing propulsion, and those who favor gradual improvements, largely finding cost savings in operational changes.

The Promise (and Price) of Air-Breathing Propulsion

Air-breathing propulsion is for many the most intuitive technology needed for advanced spaceflight. High-profile programs such as the long shelved *National Aerospace Plane (NASP)/X-30* have certainly touted the potential for exotic new high-speed air-breathing engines, such as the supersonic combustion ramjet (scramjet), to change the nature of space access. If an orbital launch vehicle could eliminate a significant part of the propellant load, such as most of the onboard oxidizer, the launch vehicle would be smaller and potentially cheaper to operate in the long term. A lower propellant mass fraction or, in other words, less propellant to lift an equivalent payload, means less mass that must be accelerated initially; overall, this further contributes to the efficiency in vehicle size.

With respect to ubiquitous military space access, air-breathing propulsion offers several beneficial characteristics. It has been suggested that among other things, a significant reduction of onboard propellant may contribute to flexibility, safety, and ultimately aircraft-like operation. Instead of fixed launch pads, many advance concepts have launch vehicles taking off from little more than long aircraft runways. Among the safety improvements cited is a potential for a controlled (and possibly powered) return to the same runway during a mission abort. It must, however, be noted that air-breathing propulsion is not necessarily a prerequisite for this type of operation; sturdy launch-vehicle construction would seem to be the only mandatory requirement for such carefree operation. Then again, a lighter vehicle at launch would definitely make such a craft easier to contemplate. Somewhat a mixed blessing, certain air-breathing engine technologies require sturdy vehicle design and construction to operate, meaning that carefree operations could be a by-product of the engine.

Several forms of air-breathing propulsion have been proposed. These range from using the atmosphere to simply augment thrust provided by rocket propulsion, to systems that use the atmosphere as a source of oxidizer. Above the atmosphere, pure rocket propulsion would be necessary, leading to criticism that in launch-vehicle applications, air-breathing engines result in significant dead weight in the form of the propulsion system or systems that can only operate for portions of the flight regime. Also, many of the air-breathing propulsion options do not function at low speeds. In response to such criticisms, many idealized air-breathing engine concepts combine multiple modes of operation in one engine. An example would have one engine operating as a self-contained rocket to get the vehicle moving from a standstill, transitioning to air breathing when conditions permit, and transitioning back to pure rocket when air-breathing operation is no longer viable, thus minimizing the dead weight problem. Alternatively, if staging is used, rocket and/or air-breathing engines may be shed when they have done their part of the flight.

Air-breathing technology does not necessarily require any technological breakthroughs, as a conventional gas-turbine-powered aircraft may be used as a "stage." The carrier or "mother ship" aircraft played a prominent role in several U.S. research aircraft programs, including several X-programs, and the lesser well known U.S. Navy Skyrocket program of the 1950s. Orbital Science's *Pegasus* launch vehicle is currently air launched from a converted *L-1011* airliner. Scaled Composites' suborbital space tourism vehicles, the X-Prize winning *Space Ship One* and the larger *Space Ship Two*, are air launched as well. Instead of being described as a first stage, the launching aircraft may also be described as a booster, or "stage 0," or simply a launching platform.

Air-launch proponents claim a much higher degree of flexibility to the space-launch system, with the carrier aircraft able to fly to the most suitable launch point. Fixed launch pads have to contend with launch windows, when orbital mechanics allow for a specific initial orbit to be achieved, and the problems of local weather conditions. This implies a degree of robustness in the launch vehicle as it must be able to cope with being carried into the air, and potentially, for long distances before being released.

Orbital launch is part velocity change to get up to orbital velocity and part achieving an altitude where orbital velocity is materially possible; therefore, the higher and faster a vehicle is already moving, the better. Lowered demands on rocket performance can equal lower cost rocket propulsion. However, higher-performance aircraft tend to be more expensive. The low-cost option of modifying existing aircraft for air launch imposes restrictions on configuration (where the launch vehicle is mounted) and maximum size. Orbital Science's *Pegasus* launch vehicle cannot be enlarged too much[11] due to the limits of available carrier aircraft, such as the converted second-hand *L-1011* wide-body airliner now in use, or the B-52[12] that was used initially.

Many launch-vehicle concepts have wings or wing-like surfaces, which will affect the flight characteristics of the carrier aircraft. Coping with these aerodynamic loads potentially leads to compromises in vehicle performance and/or expensive modifications to the carrier aircraft. Internal carriage of the launch vehicle, while avoiding many aerodynamic issues, is limited by the dimensions of the carrier aircraft's interior.[13] Release of the launch vehicle from the inside of a carrier aircraft presents additional engineering and safety challenges.[14]

The aviation company Scaled Composites on the other hand has undertaken the development of specialized aircraft for air launching spacecraft: their *White Knight* series of carrier aircraft. This effectively is the development of two complex aerospace machines: a dedicated carrier aircraft and a spacecraft. It does avoid many of the compromises needed to adapt an existing vehicle. Scaled Composites' carrier aircraft are also being marketed for roles where the payload is not released for independent flight. Pushing the boundaries of conventional aviation, respected aerospace industry magazine *Aviation Week* made claims in its March 6, 2006, cover story of a covert spacecraft launched from a large high-speed aircraft described as "XB-70 like."[15] For a brief while this article did fuel speculation and debate over this and other similar reports of large high-speed "mother ship" aircraft, but like many supposed secret aircraft being flown above the southwestern United States, it has faded from the mainstream. Without delving into the contentious claims made in the March 2006 *Aviation Week* article, the very public XB-70 program of the 1950s–1960s was a Mach 3, gas-turbine-powered aircraft,[16] and there was some discussion at the time over its potential to assist in boosting small launch vehicles.[17]

Widening the spectrum further to encompass other forms of "conventional" flight, lighter-than-air balloons have also been proposed as potential launch platforms. High-altitude balloons are a potential low-cost technology for achieving altitudes significantly higher than those possible by conventional aviation. The higher the altitude, the less atmospheric density there is to impede high-speed flight through drag and high temperatures. Balloons, while being relatively inexpensive, are generally regarded as fragile and lack the lifting potential of powered flight.

The use of conventional aviation propulsion in spaceflight for many is at best only a stepping stone, and truly revolutionary space access requires similar revolutionary air-breathing engines. This, however, runs into the many problems of processing the atmosphere for use in the engine, while the whole vehicle is accelerating through the air. A well-designed air intake will, at high speeds, assist the engine it is feeding by compressing the inbound air. In the Blackbird family of aircrafts a portion of the thrust needed to attaining Mach 3 performance is attributed to its complex variable inlet system.[18] In this regard, high performance is not simply a matter of having a powerful engine available, but one of integrating the engine with the aircraft.

The most basic jet engine is the ramjet, which dispenses with the turbine-driven compressor and uses the simple fact that the necessary air compression can be produced by having a properly configured duct and being at a high enough speed for the duct to work. It is in the details of the ducting and getting up to the necessary high speeds where this simple concept becomes an expensive and difficult-to-solve engineering problem. Both rockets and gas turbine jet engines are possible options for propelling a craft to the minimum operational speed for a ramjet. During ramjet operation these engines are either throttled down, as in the case of the Blackbird program's J58;[19] discarded, as is done in many experiments;[20] or carried along as dead weight.

Related to the ramjet is the air-augmented rocket, or ducted rocket, where the high-velocity exhaust of a rocket inside a duct draws in, compresses, and accelerates a large volume of air as additional working mass, resulting in more thrust. Like the similar ejector-lift concept for vertical takeoff aircraft, where jet exhaust is used to draw in a greater working mass, air augmentation is not a new idea, but has found little in the way of real-world application. It would appear that even with rocketry's inherent difficulties, there are more difficult aerospace propulsion technologies to put into practice. For equal performance, a big self-contained rocket is simpler to design and construct than a smaller self-contained rocket plus a duct able to provide worthwhile augmentation.

In what is defined to be a basic ramjet, the inbound air enters at supersonic speeds, but is slowed to subsonic speeds for combustion, which then leaves the engine at supersonic speeds again. This slowing, due to the air-intake configuration, results in the desired compression, but also generates waste heat and drag. A fundamental difficulty of engine/air-intake design and ramjet design in particular, is achieving worthwhile thrust after the drag of the engine itself is overcome. Eventually it no longer becomes worthwhile to decelerate the air flow any further. This of course becomes a limit to the maximum speed the ramjet can operate at. The scramjet is touted as a way to attain higher hypersonic speeds by not slowing the airflow below supersonic speeds. The challenge for the scramjet is maintaining combustion in a supersonic airflow.

Efforts to produce viable scramjet-powered vehicles have led to designs with a high degree of integration between engine and airframe. Many recent concepts involve using the entire lower fuselage in front of the engines as part of the air intake to help compress inbound air. Beyond the engine's exhaust, the airframe slopes upward, acting as an expansion ramp (essentially acting as an engine's nozzle). The simulation work that has resulted in these shapes has been greatly aided by advances in computer technology. Among the many tasks laid before the growing number of supercomputers are simulations on how the atmosphere will interact with these hypersonic aerospace craft and their engines. A computer model is, however, only a simplified version of reality, limited by the complexity of the simulation and the assumptions entered for the atmosphere, the airframe, and the engine. Up until recent physical experiments, the U.S. X-43A[21] and the Australian-led international HyShot program,[22] there were doubts as to whether a scramjet engine could sustain combustion, and whether the thrust generated could overcome the drag of the system.

Despite the principle behind a scramjet engine being sound, there remain many technological hurdles before this technology becomes practical. Closely related to the actual scramjet engine are technologies related to increasing performance margins. Among the most prominent are systems to recoup energy from the heat generated by hypersonic flight, and by the engine itself. Worthwhile acceleration by scramjet means that the vehicle will have to remain within enough atmosphere to sustain the engine but also exposing it to extreme heating. Although passive heat shielding using advanced materials is possible, many of the higher-performance scramjet concepts use active cooling to not only keep the vehicle from melting, but also to extract otherwise wasted energy. This practice is already in use in some existing rocket engines, where propellant, often a cryogenic liquid, is circulated around the engine and nozzle to keep these components from melting. However, an active cooling system adds mass; once again to be worthwhile, the energy extracted should more than compensate for the burden imposed by the cooling system.

Technical challenges also arise from the fuel being circulated itself. Hydrogen is one candidate fuel for scramjets, and is already in use in many high-performance rocket engines. By weight, liquid hydrogen provides greater energy than hydrocarbon rocket fuels such as RP-1 (a highly refined kerosene) and has no potential to leave carbon deposits or residues to clog up what must be a fast fuel flow. However, hydrogen is potentially damaging to materials due to hydrogen embrittlement—where hydrogen gets absorbed by metal alloys making it more brittle. Ironically the capacity for materials to absorb hydrogen is being investigated as a means for safe fuel storage for ground vehicles. Also, while the extreme cold of liquid hydrogen is useful for cooling, there are material problems with the temperature transition from the superheated surfaces on the outside to the cryogenic temperatures of the fuel flow on the inside. Liquid hydrogen's low density presents a drag problem due to the bulky tanks needed for this propellant. The smaller fuel tanks needed for heavier RP-1 make this less-potent fuel a viable alternative. Compounding the fuel problem for scramjet-powered vehicle concepts are the complex external shapes that airframe-to-engine integration often yields do not result in efficient internal space for fuel tanks. Among the problems plaguing the (less-challenging rocket powered) X-33 program before its cancellation was the production of a lightweight composite fuel tank able to effectively make use of the space inside the irregular lifting body shape.[23]

Thus far, air-breathing ramjet propulsion has been a prominent technology in extending the range over purely rocket-based systems. Solid-fuelled "ramrockets," essentially air-augmented rockets where the rocket exhaust contains excess fuel to be burned in the airflow from air augmentation, are being deployed in missiles such as the European *Meteor* air-to-air missile. These missiles, however, are only operating at conventional aviation speeds, and extended ranges only translate into increasing flight durations at supersonic speeds by minutes. Long duration in the atmosphere is potentially counterproductive to orbital space launch. Conventional rockets quickly clear the atmosphere to reach an altitude where orbital velocity is achievable without having to deal with severe aerodynamic drag and heating. The challenges of in-atmosphere hypersonic flight have contributed to scramjet-powered earth-to-orbit concepts, such as the long-defunct X-30 *National Aerospace Plane* switching over to rocket propulsion at lower velocities.

Avoiding some, but not all, of the problems of scramjet is the air-breathing rocket. An air-breathing rocket is basically a rocket engine that can have oxidizer supplied either by an onboard oxidizer supply, or from the atmosphere via some processing. Depending on the proposed system, air may simply be cooled and compressed, or it may actually be refined to separate out oxygen for the engines; with the remaining air components such as nitrogen and water vapor dumped overboard. In theory such an engine could operate from ground to deep space. Again to be worthwhile, the air-processing parts of an air-breathing rocket engine must be very lightweight as the mass of oxidizer it replaces must make up for the penalty of having the air-collection system along for all pure-rocket modes of flight it is hauled around with.

As long as air-breathing operations are confined to modest speeds, the air-breathing rocket does not require quite as much integration of the air-collection system with the airframe. One of the more prominent air-breathing rocket concepts being promoted today, Reaction Engine's *Skylon*, has simply mounted its air-cycle rockets in wingtip pods, leaving the main fuselage to be sturdy cylindrical shaped.[24] *Skylon* is only meant to collect oxidizer from the atmosphere up to moderate hypersonic speeds—a balance of technological risk and trade-offs its backers see as the best way to produce a cost-effective launch system in the near term.

Aircraft-Like Operations from Something Not Strictly an Aircraft

Promises of aircraft-like operations encapsulate the qualities of ubiquitous military orbital access that sets it apart from current space-launch capabilities. It also highlights that these claims have been made before. True aircraft-like operation usually translates into a lack of launch pads or other fixed infrastructure and vehicle preparation that does not involve a complete overhaul. The recently shut down space shuttle program has only been launched from the Kennedy Space Center (KSC) in Florida; with the second facility,

Space Launch Complex-6 (SLC-6) at Vandenberg Air Force Base in California, never being used for shuttle missions despite billions being spent (the shuttle launch pad there, SLC-6, being ultimately used for other launch vehicles). Preparation for a shuttle mission takes weeks, and includes carefully assembling the launch stack of orbiter, solid rocket boosters, and external propellant tank.

Every contemporary orbital launch vehicle sitting ready for liftoff is the embodiment of vast quantities of chemical energy. Harnessing this power today requires a large collection of experts and specialist personnel, leading to an expensive payroll. Sacrificing performance for inherent robustness is one potential trade-off: the cost of propellant traded for the cost of staff. Computer diagnostics and embedded sensors have made inroads in aviation, and have resulted in less maintenance hours needed to keep an aircraft fleet flying. All of these technologies and procedures that reduce personnel will of course need to be proven as effective means to safely conduct a space program and not simply corner cutting.[25]

Related to the personnel and infrastructure costs of present-day LEO access options is the problem of scheduling. Expensive as each space-launch facility is, each at best is only able to perform a few dozen launches a year. Part of the economic debate about the now-retired space shuttle program was over the high fixed and sunk costs that could not be spread out due to the low number of launches per year. This presents the interesting problem of whether low-cost orbital access depends on high flight rates, or whether only high flight rates will truly reduce the cost of orbital access.

Aircraft reusability is one quality that many spaceflight enthusiasts desire, but may not necessarily be worthwhile. Given the same technology base, a reusable launch vehicle will from a mass standpoint be less efficient than an expendable launch vehicle. The equipment needed for recovery: thermal protection, additional flight control, propulsion, parachutes, landing gear, and aerodynamic surfaces, all burden an already constrained nonpropellant mass allowance. As far as fixed costs go, the infrastructure required for vehicle inspection and refurbishment can quickly overtake any savings from reusing hardware, especially if the vehicle does not have the performance margins available to conventional aviation. Reusability becomes a valuable trait only if the price of reusability is lower than its benefits.

Aircraft-like operation implies a high degree of mission flexibility not possible with a limited number of fixed launch pads. Among the reasons for a second shuttle launch facility at Vandenburg Air Force Base during the early days of the shuttle program, was to allow for polar orbit launches not permitted from KSC on safety grounds.[26] Horizontal launch and recovery, familiar as an aircraft's runway takeoff, are also implied by the lack of a specialized launch pad.[27]

The space shuttle program was sold to the U.S. taxpayer as having an aircraft-like operation and in this regard it was a failure. The processing needed to get a shuttle off the ground and back to the launch pad has certainly made "aircraft-like" operation for the shuttle a dubious claim. To be fair, the shuttle produced was not the shuttle originally envisioned due to technology and cost trade-offs that had to be made in the 1970s when the program took shape. Today the shuttle program can be remembered for reaching far, achieving much, yet falling short on the critical goal of revolutionizing spaceflight by making it ubiquitous.

Perhaps Not All at the Same Time: Robust, Reactive, Low Cost, and History

Less exciting than totally new engine concepts, but perhaps of greater potential to transform space access, are operational changes to rocket-based launch. This takes the technology base for propelling a payload to orbit and refines it toward the goals of responsiveness, robustness, and low cost. Outright vehicle performance and sophistication is traded for robustness, reaction time, and cost effectiveness. The use of lower-performance rocketry does not mean a complete avoidance of new technologies. Instead it is more an analysis of existing and potential technologies that prioritizes cost effectiveness over outright performance. Orbital launch rocketry has a proven technology base—an expensive and difficult-to-master technology base, but the basics have been mastered by the United States and others. It only remains to be seen if this existing technology base can be redirected over time to producing ubiquitous space access.

Robust and reactive space access, though not orbital access, is already available today via pure rocket propulsion in the form of the ballistic missile. To achieve intercontinental ranges from a ballistic trajectory, these missiles generally cross all notational boundaries of space; also, due to the nature of nuclear strategy, these missiles must be able to launch within minutes of the command. Nuclear weapon miniaturization and increased missile accuracy have mitigated some of the need for increasing payload capacity; however, these factors produce two opportunities: mobile launch platforms and multiple warheads. Illustrating the robustness of modern ballistic missiles are the Russian off-road truck mobile *Topol-M,* and submarine-launched ballistic missiles (SLBMs) such as the multiple independent reentry vehicle (MIRV) equipped *Trident D-5* used by the U.S. Navy and Royal Navy.

The history of the Cold War, perceptions of threats, and the abilities of each of the superpowers to fund development and production have led to the relatively large number of long-range ballistic missiles that still exist today. Originally, ballistic-missile development was given low priority in the United States. Though the United States possessed quite a bit of rocketry

experience, including that found in the form of surrendered Nazi-rocket experts such as Werner Von Braun, it was thought that bomber technology
would suffice for the time being. This changed when it eventually became
known that the Soviet Union was not pursuing a purely symmetrical nuclear-
delivery capability in the form of bombers; instead, it had decided to heavily
invest in ballistic-missile development. In today's terms, this would be called
an "asymmetrical threat" or "game changer." Public cognition of the Soviet
rocketry threat culminated in the reaction to *Sputnik 1* being placed in orbit
in 1957. Among other things, fear of the Soviet missile and space threat led
to the creation of the Advanced Research Projects Agency (ARPA), later
renamed the Defense Advanced Research Projects Agency (DARPA).[28]
Crash programs quickly led to the first-generation U.S. intercontinental ballistic missile (ICBM), the liquid-fuelled *Atlas*, and, only a few years later, the
first U.S. SLBM, the solid-fuelled *Polaris*. Another result was, of course, the
space race.

The specifics of space launch have led to many critical differences between U.S. orbital space-launch vehicles and ballistic missiles, although
they have built from a broadly similar technology base, with several major
launch-vehicle families being direct offshoots of ballistic missiles. Weapons
development has generally favored rocketry that features robustness and reactive operation over sheer propulsive efficiency. Nuclear warheads are also
somewhat more robust than the average satellite in orbit today. Since the
retirement of the *Titan* ICBM, all long-range U.S. ballistic missiles have
been solid fuelled.[29] In general, solid-fuelled rocket motor technology is less
powerful than liquid-fuelled rocket engine technology[30] but offers characteristics important to weaponization, such as robustness and short reaction
times. Taken further is the concept of the "all-up" or "wooden" round, where
complete or near-complete munitions, such as missiles, are manufactured to
be sealed in containers for long periods of storage without maintenance, but
still can be depended on to operate with little if any preparation. This mode
of operation has made few inroads in space launch and satellite construction,
where missions are planned years in advance and both satellites and launchers are handled in continuously monitored "clean-room" environments for
as long as possible.

Proponents of ORS want to essentially develop a space operations model
that is closer to the sortie of a military unit, as opposed to an expedition. A
possible endpoint of ORS is to have stocks of launch vehicles in a state of
readiness along with payloads, or components for payloads, all of which can
be brought together on short notice. The definition of short notice is relative
to that of existing space programs, meaning that a mission timeline from ap-
proval to orbit being measured in months, as opposed to years, goes a long
way toward meeting ORS goals. Niche applications may ultimately see the

ORS concept leading to satellites attached to launch vehicles on continuous standby as is done with some nuclear weapon delivery systems.

The problem with niche applications is that their expense has to be justified in a spending environment that is often hard-pressed to sustain more routine expenditures. Arguably the continuous alert maintained for nuclear-armed ballistic missiles is a niche mission in and of itself, if only because of the extreme reluctance to employ nuclear weapons outside all but the most dire of international crises. The threat of such a conflict, and the nature of Cold War international relations, justified the expenditures of the nuclear arms race. Reactive space access, on the other hand, does not have such an easily identified or ever-present existential threat. If there was, then current technology could produce something akin to ORS, but much more expensive than that being proposed. Inventories of contemporary satellites and launch vehicles, given enough funding and personnel, could be produced and maintained in storage on the ground, ready for launch during some national emergency.

The U.S. Space Shuttle, a program both celebrated and criticized for its complexity had, in its final years, adopted something similar to a reactive mode of operation, the launch on need (LON), or rescue mission. Since the *Columbia* tragedy, space shuttle missions have been publically shadowed by preparations for a LON shuttle mission in case something goes wrong with the orbiting mission. For the most part this rescue mission made use of personnel and the space shuttle already in preparation for the next scheduled flight. However, for the final *Hubble Space Telescope* (HST) servicing mission, STS-125, the *International Space Station* (ISS) could not be used as an on-orbit refuge for the weeks it would take for the LON mission to be launched.[31] The additional dangers recognized for the crew of the space shuttle *Atlantis* on STS-125 led to some controversy over actually conducting this last servicing mission. As a precaution, the space shuttle *Endeavour* was kept on a second launch pad, ready for a quick (within a few days notice) launch during part of the *Atlantis* STS-125 mission.[32] Both the standby rescue shuttle mission and the nuclear strike mission are expensive niche applications for ostensibly undesirable circumstances, but were/are justified by the level of threat.

Necessity has led to something similar, but not the same, as ORS in the form of the Soviet military space program. To this day, versions of the Soviet R-7 ballistic missile from the 1950s continue to be used to launch payloads into orbit. With the end of the Cold War these R-7-derived "carrier rockets," the Russian nomenclature for launch vehicle, became available for Western payloads looking for low cost, but still reliable, rides to orbit. The R-7 and its derivatives use clusters of relatively small rocket engines burning kerosene, or kerosene substitutes, in liquid oxygen. These rockets are not reusable and are relatively heavy in construction. Although the Soviet Union, and now

Russia, has its share of high-performance launch-vehicle development, including many interesting concepts for hypersonic air-breathing SSTOs,[33] a combination of influences, including financial constraints and a degree of trust from decades of continuous use, have made R-7-based carrier rockets very hard to replace. The family of launch vehicles spawned from the R-7 can easily be called the "workhorse" of the Soviet and Russian space programs, and an exemplar for advocates of using simpler and cheaper mass-produced rocketry to meet expanded launch needs.

Of particular historical interest to ORS proponents was the Soviet Union's preference and sustainment of short life span satellites. Part of this was choice, and part was due to the lower technology base of the Soviet Union relative to the West. However, satellites designed to do less, and for shorter periods, did have advantages. A single launch or satellite failure has less impact, and planning for frequent replacement presents the opportunity for mass production and for frequent incremental upgrades. As in all comparisons involving the Soviet traditional emphasis on quantity over quality, concerns were raised over the vulnerability of the Western model of having fewer highly capable and long (planned) life span satellites. Now the Soviet model for running long-term space programs is not an example of the full set of ORS goals; it was a method to make up for less-sophisticated technology and not an intentional attempt to create new capabilities by increasing space access. However, the sustained higher launch rates of the past are certainly of interest.

Advances in aerospace technology that may permit the creation of the idealized air-breathing SSTO may also be applicable to robust, reactive, and low-cost multistage launch vehicles. Indeed, work on an idealized SSTO may stall its own deployment by being applied to a nearer-term multistage vehicle. Air launch from high-performance aircraft and other forms of hybrid launch vehicles, where a reusable booster is used with an orbital stage (which may or may not be reusable), can benefit from SSTO research, while at the same time being technically less risky/ambitious and less costly in the near term. The 2010 USAF *Technology Horizons* report identifies the possibility and utility of developing in the near term (2030s) a vertically launched two-state-to-orbit (TSTO) launch vehicle with a horizontally landing reusable rocket-booster stage, and an orbiter featuring air-breathing propulsion.[34] Interestingly, the same report is also advocating a very high-speed strike aircraft based on similar technology to that proposed for the air-breathing engine propelled orbiter.[35]

Pure rocketry itself has not completed all avenues for development. In general, rocket engines have high thrust-to-weight ratios and, being self-contained, avoid many of the factors that have made air-breathing propulsion technology so difficult. Already there is interest in policies to foster RP-1/LOX (liquid oxygen) rocket engine technology development and production in the United States (over continued use of Russian supplied engine technology).[36] Despite being ultimately cancelled, the X-33 program did further research on aero-spike rocket engines, a technology that promises to produce an engine nozzle that maintains high efficiency independent of altitude, instead of only being efficient at one specific altitude.

Solid rocket motors are another example of the trade-off of performance for low cost. A solid rocket motor is the combination of solid fuel and oxidizer in a binding material cast into the motor casing. It is argued that the large quantities of solid fuel needed to launch a small satellite into orbit present safety and handling problems: the rocket is always fully fuelled and therefore always capable of exploding if an accident were to occur.[37] Also, once ignited, a solid rocket motor cannot be shut down. Being fully fuelled at all times is, for missile applications, a beneficial quality once safe handling procedures are established. Recycled U.S. ballistic missiles for space launch are now exclusively solid fuelled, though often topped off with a liquid-fuelled stage or stages. Solid rocket motors were prominent in both the manned and unmanned *Ares* rockets of NASA's *Constellation* program that was cancelled in 2010. Despite a solid rocket burn through causing the *Challenger* space shuttle disaster, solid rockets have proven safe when handled correctly, as can be seen in the 100 or so shuttle missions after *Challenger*, and in the ballistic-missile fleets of the U.S. nuclear deterrent force.

The lack of controllability, and arguably safety, inherent to solid rocket motors can be overcome by replacing the solid oxidizer component of the propellant casting with a fluid oxidizer, which can be shut off or throttled if one desired this additional capability. A solid fuel combusted with a gas or liquid oxidizer forms a hybrid rocket (not to be confused with the earlier hybrid launch vehicle). Although a hybrid rocket would have a load of fuel onboard once constructed, this fuel would not easily burn without exposure to large quantities of oxidizer, which would only be loaded for launch. Hybrid rockets are in use with several manned commercial space projects, Scaled Composites' *Space Ship One* and *Space Ship Two* being two of the more prominent examples from the emerging space tourism industry.

Beyond improvements to existing types of rocketry, there is near-term potential for pulse-detonation technology, which seeks to harness the fact that detonation of fuel in many respects is more efficient than constant rate burning. Pulse detonation and other constant-volume engines also dispense with the heavy pumps or fuel pressurization needed for a high constant flow (and pressure) liquid rocket engines, potentially reducing weight further. In 2008, as part of a U.S. Air Force Research Laboratory (AFRL) program, a pulse-detonation rocket engine was operated in flight for the first time.[38] Pulse-detonation technology is also being investigated in air-breathing forms.

A potential trade-off to provide ubiquitous orbital access is flexibility. Some concepts promising to reduce launch costs, and increase launch rates, involve moving part of the mass needed for acceleration to orbital velocity off the launch vehicle. Fixed ground facilities, unlike the launch vehicle, have fewer restrictions on mass and size. The most basic would be the gun launch concept, where a very long, large bore-size gun gives a boost to a very small launch vehicle with onboard rockets. Similar in concept, but capable of boosting significantly larger launch vehicles are rail-boosted concepts, where the launch vehicle is given an initial boost while travelling on a fixed rail. Options for powering rail-based boosting range from rockets, to magnetically levitated (maglev) train technology.

Related to both pulse-detonation engines and the directed energy weapons of the next chapter are concepts of using high-power lasers or microwave beams to superheat a reaction mass. This expanding vaporized material can produce thrust in the same manner as a jet or rocket engine. The origin of this concept is credited to Dr. Arthur Kantrowitz, and the 1972 paper, "Propulsion to Orbit by Ground Based Lasers."[39] The expansion caused by sufficiently rapid heating is a detonation—the gas expands faster than its speed of sound. Rapidly cycling on and off, the external energy beam produces a series of detonations able to provide thrust. Dr. Leik Myrabo of the Rensselaer Polytechnic Institute has been prominent since the late 1990s[40] in advocating and demonstrating beamed-power pulse-detonation propulsion using the air the craft passes through as reaction mass.[41] At higher altitudes, and insufficient atmosphere, Dr. Myrabo's *Lightcraft* concept switches over to onboard reaction mass. Beamed propulsion is also under investigation globally,[42] with small models being powered upward by external laboratory lasers and microwave beams. Projecting power from a ground station to a launch vehicle in flight shares many of the same challenges faced by directed energy weapons; both involve precise transmission of large amounts of energy across hundreds of kilometers/miles of atmosphere.

Buying Off the Shelf

Technological leadership is important to national power; however, it does not necessarily have to be dependent on direct national funding. In U.S. space circles, at the time of writing, there is a great debate over the future manned spaceflight in United States, specifically over the roles of the government (NASA) and private industry. Contracts have already been signed for private companies, under Commercial Orbital Transportation Services (COTS) funding, to provide supplies to the ISS for unmanned space lifts. Many of the advanced low-cost space launchers already named in this chapter are associated with companies other than the big aerospace companies. Internet entrepreneur Elon Musk has, in only a few short years, started up a private launch company (*SpaceX*), put into service two new orbital launch vehicles (*Falcon 1* and *Falcon 9*), and constructed a recoverable space capsule for making deliveries of cargo, and potentially crews, to the *ISS* (*Dragon*). Investors are also seeing commercial viability in propulsion systems beyond the multistage rocketry used by *SpaceX* and its contemporaries. This includes the already mentioned Reaction Engines of the United Kingdom with the *Skylon* air-breathing propulsion SSTO concept.

The thresholds for military adoption and employment of a technology are not necessarily the same as that for commercial purposes. This can already be seen with information technology. In many respects, common-day civilian computer and information applications display a level of sophistication that is not seen in general military use. Making the trade publications recently was a U.S. military program investigating the value of smart phones.[43] Smart phones in the civilian world are ubiquitous items and are cited by some as a major factor in contemporary Western culture. However, the relative fragility, both in physical and software terms, of consumer electronics has limited military applications. The commercial failure of the Concord is an example of the opposite; although the military has found the capability for supersonic flight to be a routine requirement for many categories of combat aircraft, this technology base faces greater challenges to being applied to commercial flight. It remains to be seen if the technologies behind the slowly emerging industry of civilian suborbital and orbital access may be suitable for national defense needs.

It should, however, be remembered that for commercial manned spaceflight to become a viable industry beyond serving a small clientele of wealthy adventurers, the cost and timelines of a space launch would have to come down significantly, and reliability would certainly have to increase. Ubiquitous civilian/commercial orbital access may involve technological hurdles beyond that needed for initial fielding of a military form of ubiquitous orbital access. The fact that there are companies out there working to solve the greater problems of commercial manned spaceflight leave open the possibility of ubiquitous military spaceflight being spun-on from the civilian marketplace, instead of civilian spaceflight being a spin-off of military technology. It also raises the specter that lack of attention in the area of civil launch-vehicle development will allow a potential opposition force (OPFOR) to reap the benefits of commercially viable private spaceflight.

Opening Doors

Ubiquitous LEO access will affect the options that are available with respect to the exploitation of space. It is consistently reinforced that the conventional-warfare preeminence enjoyed by U.S. military today is linked to its use of space-borne assets; therefore, ubiquitous LEO access will affect today's conventional-warfare paradigms. The ongoing proliferation of space-access technology has the potential to accelerate if ubiquitous LEO access becomes a reality. This proliferation will affect the nature and number of threats to the Western world in general as well as specifically to the United States.

There are longstanding warnings, such as those found in the 2001 Rumsfeld Commission report, concerning the long-term viability of the existing model for U.S. military space supremacy. Recognition of the force enhancement conferred by the long investment in space by the United States has spawned two areas of threat: attempts to emulate U.S. military space power, and attempts to neutralize U.S. military space power. The high cost of LEO access has so far limited the ability of all but near-peer competitors to field such strategies. Near peers such as the People's Republic of China and Russia, possess high-technology economies and military industries, meaning they have the option of developing equivalents to technology that regulations prevent them from importing (legal or otherwise). Even without espionage providing shortcuts, once a basic concept is proven, the research and development of equivalent and counter-technologies is only a matter of time. Potentially, the tech bases of a near peer, or even of traditionally allied nations, could introduce disruptive technology in the form of ubiquitous space access.

Added to the expected threat provided by competing foreign military space programs are concerns about commercial space services as both a threat and as vulnerability. Commercial entities are now providing space services, such as earth imagery, that only a few years ago would have been the exclusive domain of national agencies. Foreign companies that provide these services are beyond the reach of U.S. "shutter control" regulations that U.S. companies are subject to. Among recent operations, the United States bought exclusive rights to all commercial imagery of the Afghanistan theater of war to ensure operational security—a step it had also taken during the Gulf War. These financial methods may not always be available, or effective. Increased use of commercial communication satellites to handle the seemingly endless growth of U.S. military data bandwidth needs has become a necessary stand-in for delayed military programs. In general, the military use of commercial satellites has the potential to mark these satellites as targets in any future conflict.

Door 1: Continuation of Sanctuary (a.k.a. Survivability)

The academic world often seeks to sort the real world into theoretical boxes or to arrange them into a theoretical spectrum.[44] Among the more commonly cited works is USAF Lieutenant Colonel David E. Lupton's *On Space Warfare*, which divides military space policy and doctrinal positions into: sanctuary, survivability, high ground, and space control.[45] The term high ground in particular is often used, with reverence, to describe how far a nation could go with military space power, and defines space as being the critical domain with respect to not just one specific battlefield or conflict on earth, but to a nation's global destiny (see also Evertt Dolman's *Astropolitik*). The value of specific capabilities and policy doctrines associated with overt space weaponization fluctuate with the price of LEO access. Therefore, the emergence of ubiquitous space launch, and the form it takes, will shape what aspects of these doctrines become viable.

Space as a sanctuary is not a purely pacifist stance, instead it is also a strategy to use diplomacy to safeguard the existing security situation for Western military power. During the Cold War, and to the present day, satellites were part of the nuclear deterrence strategy. Early warning satellites gave adequate notice to ensure a retaliatory strike, and spy satellites (national technical means) promoted stability by first confirming the existence of retaliatory capability, dispelling the temptations of a first strike, and as a confidence-building measure for treaty verification, dispelling the temptation to cheat. Space served as an environment to promote strategic stability in the Cold War. In the age of space-force enhancement, some Western nations, such as Canada, are wary of active space weaponization due to the threat it may bring back on the West. The West and the United States, in particular, have heavy investment in space infrastructure; today space is arguably not just an out-of-mind pillar for Western military power, but also for Western society in general. Preventing the development of space weapons through diplomacy safeguards this pillar. This, however, is a stance that has been under contention for a while, even with space access having a high-cost barrier.

With the introduction of ubiquitous space access, whether through direct military investment or not, space sanctuary, as a policy, may no longer be sustainable. For one thing, the space sanctuary position does not freeze the situation to the West's advantage, allowing others to catch up, something near peers such as China and Russia are openly doing. That is not to say that U.S. force-enhancement capabilities cannot improve, but with precision munitions already having accuracy within single digit measurements in meters, and charges that communications bandwidth is being flaunted for unnecessary multimedia presentations and excessive micromanagement (as opposed to more effective forms of support for front line troops), one can see diminishing returns from purely advancing space-force-enhancement capabilities. Increased access to LEO, unless it is somehow exclusive to the United States, would actually in this regard, threaten U.S. power by facilitating others in replicating U.S. space power capabilities.

In recognition of U.S. satellite vulnerabilities and emerging foreign antisatellite (ASAT) capabilities there have been recent calls to reduce dependency on space assets as much as possible. However, again it should be remembered that space assets provide the ability to project information superiority into denied parts of the world. Compact satellite telephony services such as *Iridium* may have been a mismatch for the civilian market; however, its global reach and compact handset (relative to other satellite communications gear) made it ideal for many power projection applications. As long as the United States wishes to retain the capacity to project power globally, and continue to fight with the level of precision to which the West has become accustomed, then some amount of space infrastructure will be needed.

Short of abandoning assured use of space, but without proceeding with a policy of clear and definitive space weaponization, ubiquitous LEO access provides options to quickly reconstitute military space assets in times of war—a strategy of survivability. As noted earlier, this strategy could be put into place today were it not for the extreme cost associated with present-day orbital launchers. A ubiquitous space-launch capability combined with low cost but capable small satellites (see Appendix) would assure U.S. space power through sheer numbers of easily replaced and disposable satellites. It is unlikely that another nation would attack U.S. space assets as a goal in itself. Instead, any attack on U.S. military space infrastructure would be performed as a means of hindering or neutralizing U.S. forces on earth. A recognized capability for U.S. space power to be quickly rebuilt as part of a swift and devastating retaliation would, then, perhaps be deterrence against an ASAT attack in the first place, and possibly even deterrence against whatever activities the opposition force is planning on earth. This would, however, also require maintaining war stocks of critical satellites.

The ability to rapidly reconstitute existing capabilities is only the beginning. With greater access to orbit, there is also the potential to surge space capabilities on demand. Many space assets that could be useful in times of war may not be economical to maintain in orbit in times of peace. The shortened timelines promised by ORS advocates raise the possibility that ingenious new uses of space could be thought of and implemented in the timeframe of a single conventional war. Moreover satellites in orbit, even without open warfare, are continuously exposed to the hostile environment of space, where radiation is plentiful, collisions often involve high energies, and when closer to earth, what little atmosphere is present includes highly reactive single oxygen atoms.

Aside from the high cost of space launch, the exploitation of space (let alone the exploration of space) is expensive due to the challenges of surviving in a hostile environment. The high cost of a space mission today is compounded by the need for engineering perfection due to the very limited capacity for on-orbit tweaking of equipment and repair, which is, for most missions, out of the question. In an odd way, budgeting for loss and failure is a luxury that contemporary use of space cannot enjoy. Keeping space assets on the ground, a space force in potential, while training with earth-bound substitutes and simulations to maintain credibility, could be a low-cost means of maintaining U.S. space power.

Not only do orbital space weapons become more affordable due to effective reductions to launch costs, but the potential targets, both United States and potential opposition force satellites, also increase due to access. Today's satellites are considered strategic assets, and certainly have price tags in the same range as warships. The loss of one of today's militarily critical satellites would be akin to the loss of a major national asset and would demand an appropriate response. Indeed at some stages of the Cold War, an attack on specific satellites, early warning and strategic communications in particular, could have been the beginning of a strategic nuclear war. As ORS and related concepts bring satellite assets to the "tactical" level, through lower cost and greater numbers, their abundance raises the possibility that at least some of these new military space assets would be regarded as being more expendable. Lowered importance creates the problem of an adequate response to a satellite attack, especially now in an age where Western militaries seem to be increasingly constrained by requirements for proportional responses. In other words, if satellites come to be regarded as tactical assets, how much escalation can be justified over the loss of a few of these "tactical satellites?"

With escalation, satellite services that are considered critical today come under threat. Once again there is the problem of defense or deterrence. Despite being the low cost and existing solution, there are political barriers to threatening nuclear retaliation for the loss of major space infrastructure—even if it is a pillar of the Western way of war and perhaps even Western society. Like the moral dilemma to nuclear deterrence, solutions range from diplomacy to armaments.

Door 2: Space Weaponization

There are many potential futures for the military use of space. To a large degree, these futures are differentiated by the level that space becomes weaponized. At present the existence or nonexistence of space weapons is based largely on what one defines a space weapon to be. Despite actually destroying orbiting satellites, direct ascent (earth-to-space) systems such as the Chinese satellite interceptor that conducted a destructive test in 2007, the U.S. Air-Launched Miniature Homing Vehicle (ALMHV) tested in the 1980s, and the

U.S. SM-3 missile system adapted for a satellite interception in 2008, would not be considered space weapons under the space weapons ban recently proposed by China and Russia in the forum of the United Nations (UN) sponsored Prevention of an Arms Race in Outer Space (PAROS), where space weapons are defined as orbiting weapons and not satellite-destroying weapons.[46] It should be noted that the Chinese government's own direct ascent testing did not stop them from protesting the 2008 U.S. satellite shoot down, which was less a weapons test and more of an action to mitigate the threat of a chemical accident if the failed U.S. satellite had been allowed to crash intact.[47] The present state of space weaponization is arguably unclear and somewhat "fuzzy."

Historically, a space weapon defined as an orbiting weapon has been deployed, though only temporarily. The Soviet Union mounted cannon armament on at least one of the *Almaz* series of military space stations, which were covertly orbited as members of the civilian Salyut program.[48] The *1967 OST* does not ban conventional weapons in orbit, and disguising the *Almaz* program had more to do with it being a reconnaissance (spy) platform. The ease at which a satellite's true purpose may be disguised presents serious problems to treaty verification. Indeed the whole subject of satellite "inspection" becomes problematic due to the weapons-related nature of this capability. An orbiting space weapons ban could ironically ban the means to verify compliance of its own terms.

Among ground-based systems that are often overlooked as space weapons are electronic warfare (EW) systems meant to disrupt receiving services broadcast by satellites. GPS jammers were deployed by Iraq in a failed bid to interfere with U.S. precision munitions during the 2003 war.[49] The U.S. *Counter Communication System* (*CounterCom*) is listed in publically available budget documents under the category of "Counterspace Systems."[50] Now EW is often overlooked in general: it is kept secret intentionally as it is related to signals intelligence and other forms of espionage, and the lack of destructive (or "kinetic" despite kinetic meaning something else in physics) effect leads to a degree of natural obscurity to the public, and possibly even policy makers. Now *CounterCom* is effectively a portable satellite communications station, and by all descriptions has a generally benign appearance. An on-orbit EW system on the other hand may well garner much more attention.[51]

Less clear examples of potential notational space weapons begin with long-range ballistic missiles, where range is achieved by having ballistic courses that cross not only all notational boundaries of space, but also go beyond some definitions of LEO up to some medium orbits. If orbiting is important, then one cannot leave out the Soviet Union's fractional orbit bombardment system (FOBS), a nuclear weapon delivery system deployed for a short time toward the end of the Cold War. To this day, the "legality" of FOBS, under the *1967 OST*, which specifically bans orbiting nuclear weapons,[52] is the subject of academic debate.[53]

Another potential candidate for inclusion as a destructive space system would be ground- and air-launched weapons that utilize satellite support for their accuracy in hitting fixed ground targets. Weapons such as bombs equipped with the Joint Direct Attack Munition (JDAM) system are satellite assisted; these exhibit very different levels of accuracy when based purely on inertial guidance vis-à-vis inertial guidance with GPS assistance. For the JDAM weapon system, this is a dramatic reduction of circular error probability (CEP), a measure of accuracy, from 30 meters on inertial guidance down to only 5 meters when GPS is available.[54]

Combining all of the earlier mentioned aspects is the Chinese antiship ballistic missile (ASBM). This is a long-range land-based naval weapon thought to be based on the DongFeng-21 (DF-21), or CSS-5 under NATO nomenclature, ballistic missile.[55] The ballistic missile is used to deliver a warhead capable of terminal guidance and maneuvering to hit a mobile seaborne target. Interestingly there is speculation that other offshoots of the DF-21 program include the booster thought to have been used to deploy the ASAT used in the destructive 2007 test, the booster for the Chinese anti–ballistic missile (ABM) program, and a commercial solid-fuelled launch vehicle.[56] The solid-fuelled DF-21 medium-range ballistic missile is known to be road mobile, appearing occasionally in military parades. The lack of transparency by the People's Republic of China on military matters in general also shrouds the specifics of these four programs, though both the ASAT test[57] and the later ABM test in 2010[58] were eventually acknowledged by Chinese government sources. In comparison, the United States has been publically debating the value of ASAT weapons, missile defense, and prompt global strike via converted ballistic missiles—all systems that utilize large solid rocket motors, as do U.S. commercial launchers based on surplus ICBMs.

There is also the matter of ad hoc weaponization, where a spacecraft capable of maneuver is used to cause damage, but is otherwise not specifically designed to be a weapon. This group includes the Canadian-designed shuttle remote manipulator system (RMS) or *Canadarm*, which did figure as part of Soviet fears about the U.S. Space Shuttle program. Earlier U.S. and Soviet space weapons research had plans for military crews aboard existing space capsules rendezvousing with target satellites for inspection and neutralization. To this day the phrase "satellite inspection," continues to raise alarm in segments of the disarmament community.[59] Essentially any on-orbit rendezvous capability, no matter how benign its intentions are, can be construed into a weapon. Among the more worrisome of "benign" concepts are space tugs for moving large satellites into mission orbits, extending mission life when onboard propellant is exhausted, and removing satellites from mission orbits once a satellite is at the end of its life. Moving a satellite unwillingly from its mission orbit is an on-orbit attack without any of the problems of debris.

The unclear nature of what constitutes a space weapon also presents a case to be made that the mechanism of coercive force (in space) is less important to widespread weaponization of space than access to space itself. Space as a sanctuary from warfare is largely by "default," a product not purely of pacifist intentions, but instead of budget and strategic realities for all actors on the international stage. Dropping a digit or two of the cost per pound or kilogram to orbit will therefore enable not just the major powers to actively threaten each other's orbital assets, but potentially open the door for other forms of conflict in space. Commercial dispute over one of the limited number of lucrative geostationary orbital slots has resulted in what can be described as belligerent acts between satellite operators, as was the case with *Palapa B1*. The stakes are higher for nationally owned military satellites. These assets are in the same price range as major warships, as is arguably their importance as a military asset; this perhaps lends an additional level of restraint. Low-cost access to orbit may effectively lower the value of individual space assets, and therefore perhaps lower the restraint-inhibiting military action. The specifics of the "military action" are limited only by the laws of physics and imagination, once the price is right of course.

Though much has been said about the specific wording of the space policies of individual U.S. presidencies, there have also been significant continuations of policy across several decades concerning the prospect of space weapons. Although U.S. counter-satellite (counter space) activities have so far been modest (the *CounterCom* system for instance only has a budget measured in millions of dollars[60]) the policies for several presidencies have consistently included statements related to maintaining freedom of action in space for United States and allies, and the capability to deny or disrupt the space capabilities of threats. As with past presidents, the space policy of the Obama administration continues the U.S. stance against arms control measures that "impair the rights of the United States to conduct research, development, testing, and operations or other activities in space for U.S. national interest."[61]

Proliferation and Arms Racing

Space access is only proliferating, and almost by definition, ubiquitous space access implies a degree of international access. Easy international access to space means potential for greatly increased international competition. There is already growing competition today in military space applications even without easy access to orbit. Despite the high cost of establishing the large constellation of satellites necessary for a global positioning service, and despite the United States offering reasonably accurate and reliable access to its own NavStar GPS as a "global public good," there are several competing systems emerging: the European Union *Galileo* system slowly being orbited, Russia's reconstituted Soviet era *GLONASS*, and China's current ambitions to expand its regional positioning system into a global system. These independent satellite navigation systems are already a point of concern, if not outright alarm, for the U.S. defense community. However, for their backers, the lack of a navigation system independent of U.S. control is a source of concern and alarm. In the case of competing navigation systems, there have been efforts among some of the technical groups involved to avoid potential signal interference, and recognition of the potential unwanted consequences of this competition. However, not all potential space competitions may involve such agreement over the need for consultation and coordination.

Low-cost space access is a disruptive military technology, with the potential to change the perceptions that have limited the weaponization of space. As with many introductions of disruptive technologies there is potential for an arms race to develop. One nation's attempt to seize the initiative, or at least build up a defense, triggers another nation to embark on a competing weapons program. This second nation's competing program is interpreted as a threat by the first nation, necessitating a response in the form of increases to its own weapons program. The second nation, on seeing another increase in the weapons program of the first nation, is pressured to do the same, leading to an escalating cycle. Ubiquitous space access may make the initial stages of a space weapons arms race affordable. Once a space arms race develops, then for a while at least there is increased tolerance for the expense such weapons may bring.

Among nonpacifist arguments against space weapons development, is the point that initiation of overt space weaponization will not only start an arms race, but also an arms race on a nearly level playing field. The counterargument is that the success of U.S. space-force enhancement over the last two decades itself is cause for space weaponization. Not only are others emulating U.S. space-enhancement capabilities, potentially creating targets for future U.S. counter-space systems, but others are also already looking into countermeasures to existing U.S. space enhancement. The Chinese GPS plans provide a possible case of the former, and the Chinese ASAT program is an example potentially of the latter.[62] Without active hostilities, however, all this remains in the theory and potential capability stage.

If outer space is the highest of the high ground, as it is often regarded, there will always be a military or security component to space activities. As long as space is not an overtly weaponized environment there will be pressure to do so. Although comparisons are drawn between space and other seemingly weapons-free environments, such as Antarctica and the depths of the seabed, it must be remembered that at present there is little human activity, let alone military interest, in Antarctica and the harsh depths of the sea. Orbit, on the other hand, has been a military and strategic concern from the very beginning of the Space Age. Relaxed access to the earth's orbit can only bring more pressure on a nation, any nation with a comparable high technology base as the United States, to be the first to overtly weaponize space.

Conclusion

Over the history of launch-vehicle development, techniques to deliver performance gains have faced real-world engineering problems that often erase any theoretical bonuses. Often this has been a result of technology simply not being mature enough for real-world use, or unforeseen problems with integrating many new and promising technologies into a solution. This has led to debate between those who back investment in revolutionary but risky technologies versus incremental changes to multistage launch vehicles that, to a large extent, still depend on pure rocket propulsion to reach orbit. On one end of the mainstream there is the reusable air-breathing engine powered SSTO launch vehicle. On the other end is the refinement of multistage rocket-powered launch-vehicle technology into something that is robust, can be launched on demand, and procured cheaply enough to greatly expand the possible uses of space.

Notes

1. Global Security, "Advanced Extremely High Frequency (AEHF)," http://www.globalsecurity.org/space/systems/aehf.htm

2. Lockheed Martin, "Advanced Extremely High Frequency (AEHF)," http://www.lockheedmartin.com/us/products/advanced-extremely-high-frequency—aehf-.html.

3. United States Air Force, "Factsheets: Advanced Extremely High Frequency (AEHF) System," http://www.losangeles.af.mil/library/factsheets/factsheet.asp?id=5319.

4. Sheila Rupp, "Operationally Responsive Space," *Air Force Print News,* May 22, 2007, http://www.kirtland.af.mil/news/story.asp?id=123054292.

5. The term suborbital is also used sometimes, but suborbital trajectories can and do reach altitudes well above LEO altitudes.

6. The principal atmosphere science layers are usually given as: troposphere, stratosphere, mesosphere, thermosphere, and exosphere. Other layers in the vertical realm important to specific earth sciences include the ozone layer and ionosphere.

7. National Museum of the U.S. Air Force, "Fact Sheet: Lockheed B-71 (SR-71)," http://www.nationalmuseum.af.mil/factsheets/factsheet.asp?id=2699.

8. James Oberg, "Astronaut." *World Book Online Reference Center.* 2005. Chicago, IL: World Book, Inc., http://www.worldbookonline.com/wb/Article?id=ar034800.

9. A Mach number is the speed of an object divided by the speed of sound in that medium. With no medium for the transmission of sound in outer space, this usage of Mach generally relates to a comparable speed within the atmosphere under specific pressure and temperatures.

10. Indeed some engineering studies and literature, such as Andrew J. Butrica's *Single Stage to Orbit: Politics, Space Technology, and the Quest for Reusable Rocketry*, claim the high performance of the *Saturn* V third stage, the one responsible for the lunar trajectory, could have allowed adaption of the *Saturn* V third stage into an expendable SSTO launch vehicle.

11. For larger payloads, *Pegasus* stages are stacked on top of another rocket stage, for launch pad launch, forming the Taurus I launch vehicle.

12. Early *Pegasus* launches were from a NASA B-52. According to NASA public relations material, the same B-52, under NASA tail number 008, was used as mother ship for many experimental high-speed aircraft, including the X-15 program, which earned a handful of test pilots astronaut status.

13. Internal carriage was recently explored by Air Launch LLC's *Quick Reach* launch vehicle under funding from DARPA's Force Application and Launch from CONUS (FALCON)—Small Launch-Vehicle program.

14. Marti Sarigul-Klinjn, et al., "Trade Studies for Air Launching a Small Launch Vehicle from a Cargo Aircraft" (Reston, VA: [0] American Institute of Aeronautics and Astronautics, 2005), http://www.airlaunchllc.com/AIAA-2005–0621.pdf.

15. William B. Scott, "Two-Stage-to-Orbit 'Blackstar' System Shelved at Groom Lake?" *Aviation Week,* March 6, 2006, www.aviationweek.com.

16. The XB-70 is the largest Mach 3 capable aircraft of making it into the sky, if only for a fleeting few experimental flights in the 1960s.

17. William G. Holder and William D. Siuru, Jr., Captain USAF. "Some Thoughts on Reusable Launch Vehicles," *Air University Review,* November-December 1970, http://www.airpower.au.af.mil/airchronicles/aureview/1970/nov-dec/holder.html.

18. See also former Lockheed Skunk Work's head Ben Rich's account of Blackbird development in his 1994 book *Skunk Works,* coauthored with Leo Janos, which includes his involvement with the propulsion system as an engineer under the first Skunk Works leader, Kelly Johnson.

19. The high proportion of air that bypasses the turbine sections of the J58 engines to be fed directly into the afterburners has the old Blackbird family of aircraft's propulsion system sometimes described as a turbo-ramjet.

20. Experimental ramjets have been launched on top of multistage expendable boosters, including Orbital Science's *Pegasus*.

21. NASA, "NASA's X-43A Scramjet Breaks Speed Record," November 16, 2004, http://www.nasa.gov/missions/research/x43_schedule.html.

22. University of Queensland, "Hyshot," http://www.uq.edu.au/hypersonics/? page=19501.

23. NASA, March 1, 2001, Press Release, http://www.nasa.gov/home/hqnews/ 2001/01–031.txt.

24. Reaction Engines LTD., "Current Projects: SKYLON," http://www.reac tionengines.co.uk/skylon.html.

25. There are proposals for some bulk low-cost payloads such as propellant supplies for on orbit refueling, where a high failure rate is acceptable as long as the cost of the lost payload does not surpass the savings from mass-produced cheap "dumb" boosters. These proposals would also minimize the cost of launch insurance as well by simply accepting the cost of lost payload as a part of the business model.

26. A polar launch from KSC would involve a trajectory passing over, and potentially dropping boosters and other debris on populated areas.

27. There is nothing in the nomenclature of aircraft-like operation to rule out a vertical launch and recovery; however, to avoid excessive ground damage, launch pads are equipped with various means to redirect and diffuse the sound, pressure, and heat that result from a rocket launch.

28. Except for a brief period in the 1990s, when DARPA went back to the Advanced Research Projects Agency (ARPA) name.

29. The Soviet Union persisted with storable liquid-fuelled rocket engines in their land- and submarine-based ballistic missiles up to the end of the Cold War.

30. Solid rockets use the term "motors," whereas liquid-fuelled rockets use the term "engines."

31. The *ISS* and *HST* have very different orbits, making it impossible for a single space shuttle mission to visit both.

32. NASA, "STS-400: Ready and Waiting," May 5, 2009, http://www.nasa.gov/audience/foreducators/sts400-ready-and-waiting.html.

33. Among published Russian concepts is the Russian Hypersonic System Research Institute's AJAX, an aerospace craft/propulsion system that uses electromagnetic techniques to overcome energy losses inherent in scramjet operation.

34. United States Air Force, *Report on Technology Horizons: A Vision for Air Force Science & Technology During 2010–2030 Volume 1*, May 15, 2010, http://www.af.mil/shared/media/document/AFD-100727-053.pdf.

35. Ibid.

36. Brian Berger and Amy Klamper, "NASA Propulsion Plans Resonate with Some in Rocket Industry," *Space News*, February 26, 2010, http://www.spacenews.com/launch/100226-nasa-propulsion-plans-resonate-rocket-industry.html.

37. Brazil's indigenous launch-vehicle program suffered such a loss in 2003 when a prototype *Veículo Lançador de Satélites* (VLS), an all-solid-fuel orbital launch vehicle, exploded days before launch, killing several ground crew and destroying the launch pad.

38. Larine Barr, 88th Air Base Wing Public Affairs, United States Air Force, "Pulsed Detonation Engine Flies into History," *Air Force Print News Today*, May 16, 2008, http://www.afmc.af.mil/news/story_print.asp?id=123098900.

39. American Institute of Beamed Energy Propulsion, Inc., http://www.aibep.org/Kantrowitz.htm.

40. Leik N. Myrabo and Donald G. Messittf, "Ground and Flight Tests of a Laser Propelled Vehicle," 1997, http://pdf.aiaa.org/downloads/1998/1998_1001.pdf?CFID=1326408&CFTOKEN=88988801&.

41. Brittany Sauser, "Riding an Energy Beam to Space," *Technology Review*, August 5, 2009, http://www.technologyreview.com/blog/deltav/23928/.

42. Christopher Mims, "Microwave-Powered Rocket Ascends without Fuel," *Technology Review*, September 7, 2010, http://www.technologyreview.com/blog/mimssbits/25701/.

43. U.S. Army, "Connecting Soldiers to Digital Applications." *Stand-To!*, July 15, 2010, http://www.army.mil/standto/archive/2010/07/15/.

44. Mueller, Karl P., "Totem and Taboo: Depolarizing the Space Weaponization Debate." Paper based on presentation given to Weaponization of Space Project of the Eliot School of International Affairs Space Policy Institute and Security Policy Studies Program, George Washington University, December 3, 2001, http://www.gwu.edu/~spi/spaceforum/TotemandTabooGWUpaperRevised%5B1%5D.pdf.

45. David E. Lupton, Lieutenant Colonel, USAF. *On Space Warfare*, 1998, http://www.dtic.mil/cgi-bin/GetTRDoc?Location=U2&doc=GetTRDoc.pdf&AD=ADA421942.

46. "Draft Treaty for the Prevention of Placement of Weapons in Outer Space," February 12, 2008, http://www.ln.mid.ru/brp_4.nsf/e78a48070f128a7b43256999005bcbb3/0d6e0c64d34f8cfac32573ee002d082a?OpenDocument.

47. Tom Bowman, "China Protests after U.S. Shoots Down Satellite," *National Public Radio*, February 21, 2008, http://www.npr.org/templates/story/story.php?storyId=19246330.

48. David S.F. Portree, "NASA, Mir Hardware Heritage," March 1995, http://ston.jsc.nasa.gov/collections/TRS/_techrep/RP1357.pdf.

49. Jim Garamone, American Forces Press Service, "CENTCOM Charts Operation Iraqi Freedom Progress," March 25, 2003, http://www.defenselink.mil/news/newsarticle.aspx?id=29230.

50. United States Air Force, "Program Elements FY2009—Counterspace Systems," February 2008, http://www.js.pentagon.mil/descriptivesum/Y2009/AirForce/0604421F.pdf.

51. The commercial satellite industry is already well aware of the effects of on-orbit signals interference. In the 1990s there were the exploits of *Palapa B1*, an Indonesian-owned commercial communication satellite, at the center of several disputes over the use of a geostationary orbital slot claimed by the island nation of Tonga and leased to third-party satellite operators. Notable belligerent acts include this satellite and another satellite hazardously attempting to occupy the same orbital slot in 1992, and accusations of radio-frequency interference (jamming) with another satellite it was in dispute with in 1996. More recently in 2010 another large GEO communications satellite, *Galaxy-15*, began drifting, due to suspected damaging from natural solar activity, but inexplicably was transmitting on its own for several months, despite efforts of ground control to shut it down, earning to the chagrin of its owners the nickname of "Zombie-sat." Several other satellites serving North America had to be repositioned to avoid *Galaxy-15* and the signals interference it was causing as it strayed into the orbital slots of working satellites in the GEO belt.

52. *1967 OST*.

53. Instead of lobbing multiple nuclear weapons on a high ballistic arc, FOBS used a variation on a Soviet "heavy" ICBM to loft a nuclear warhead into a very low orbit. Before completing one orbit, the warhead would reenter the atmosphere

to hit a surface target. An orbital flight path allowed the warhead to approach from any direction, allowing it to attack through gaps then present in U.S. ballistic missile early warning system.

54. United States Air Force, "Factsheet: JOINT DIRECT ATTACK MUNITION GBU-31/32/38," November 2007, http://www.af.mil/factsheets/factsheet.asp?id=108.

55. Department of Defense, *Military and Security Developments Involving the People's Republic of China 2010*, http://www.defense.gov/pubs/pdfs/2010_CMPR_Final.pdf.

56. Global Security, "DF-21/CSS-5," http://www.globalsecurity.org/wmd/world/china/df-21.htm.

57. Shirley Kan, *China's Anti-Satellite Weapon Test*, Congressional Research Service Report for Congress, April 23, 2007, http://www.dtic.mil/cgi-bin/GetTRDoc?AD=ADA468025&Location=U2&doc=GetTRDoc.pdf.

58. Russell Hsiao, "Aims and Motives of China's Recent Missile Defense Test," *China Brief* 10(2), January 21, 2010, http://www.jamestown.org/single/?no_cache=1&tx_ttnews[tt_news]=35943.

59. Wilson W.S. Wong and James Fergusson, *Military Space Power* (Santa Barbara: Praeger, 2010), 91.

60. United States Air Force, "Program Elements FY2009—Counterspace Systems," February 2008, http://www.js.pentagon.mil/descriptivesum/Y2009/AirForce/0604421F.pdf.

61. *National Space Policy of the United States of America*, June 28, 2010, http://www.whitehouse.gov/sites/default/files/national_space_policy_6-28-10.pdf.

62. Department of Defense, *Military and Security Developments Involving the People's Republic of China 2010*, http://www.defense.gov/pubs/pdfs/2010_CMPR_Final.pdf.

CHAPTER 3

Directed Energy Weapons

Directed energy weapons (DEWs), the ability to project destructive energy without the need for a physical projectile, has been on the emerging technology horizon for decades, if not since antiquity. There is persistent debate over whether Archimedes of Syracuse (287–212 BC) constructed a solar-focusing weapon. H. G. Wells's 1898 science fiction classic, *The War of the Worlds*, had mention of heat rays employed by the technologically superior invaders from Mars. During the interwar period, early radar work in the United Kingdom was linked to research over the feasibility of a radio "death ray." In the 1980s, various DEW concepts were investigated under the Strategic Defense Initiative (SDI), a comprehensive plan to find a defense against large-scale intercontinental ballistic missile (ICBM) attack. Recent efforts include goals such as shooting down rocket, artillery, and mortars (RAM) threats and igniting improvised explosive devices (IEDs). Though lacking the ambition of past DEW concepts, the capabilities being demonstrated today are perhaps finally heralding the transition of DEW from a promising emerging military technology to practical battlefield systems.

Along with beams of fantastical destructive power, DEW technology also promises less-than-lethal coercive effects. There has been a growing desire to give soldiers and law enforcement a stun weapon, something common in science fiction. Many direct energy weapon concepts include variable effects, allowing operators to tailor effects to only what is appropriate to a particular situation. A component of the last Revolution in Military Affairs was increased precision, lowering the investment in force and collateral damage needed to accomplish a task. DEW technology has the potential to take precision to new levels, the ability to not just target individual troublemakers in a crowd, but to apply only enough force to change their actions. Now although this clean and bloodless vision may seem overly optimistic, the promise of this capability does fit into some perceptions and desires over how military force can and should be used by the West.

Although DEWs promise much, the inescapable reality is that these concepts have been around for quite some time without much in the way of tangible products. The problems can be divided into categories based on the components of the term DEW—there are challenges to reliably direct the energies involved, challenges to generating the energies in the correct form, and challenges in producing effects on the target to make these systems actually weapons. Technical challenges translate into large planned, and many unplanned, financial costs for programs seeking to create a practical DEW. The perceived newness of DEW technology, and skepticism brought on by awareness of past failures, present additional roadblocks to these systems being accepted. DEWs may represent a redundant capability that, outside of a few niche applications, has quite far to go before matching bullets and warheads in general military utility. With the severe budgetary constraints faced by the United States and others in the second decade of the 21st century, there will be great competition, and outright opposition from within the defense establishment.

There is also resistance brought on by concerns that these systems may have unintended or undisclosed effects; in particular, concerns over the potential for less-than-lethal DEW capabilities to be abused. A weapon that does not kill to achieve effectiveness is not only useful in the prevention of unnecessary death, but the capacity of DEW in the wrong hands to torture and maim also presents serious concerns over proliferation to nations with questionable human rights records. Accidental misuse could also stem from overuse due to less-than-lethal weapons not being handled with the same caution and gravity as traditional killing weapons. Perceptions on the "legality," and sentiments on how "humane" a weapon is or should be, may overshadow any real obligations a nation may have concerning a particular weapon system. Rightly or wrongly weapon concepts do acquire stigmas that can limit any opportunities they have for demonstrating their utility, including that of preventing unnecessary casualties, on real battlefields.

Energy and Energy Weapons

Energy is the capacity to do work, to produce change in a physical system. An energy weapon is a weapon that performs some manner of physical change, considered of a destructive nature, without the benefit of a physical projectile or container. The cliché energy weapons portrayed in the media are those that transmit destructive powers via electromagnetic (EM) radiation, usually in the form of beams of visible light. In reality, most proposed weapons based on EM radiation use parts of the spectrum invisible to the human eye. Indeed most of the EM spectrum is not perceivable by the human eye, ranging from many-meters-long wavelength radio waves, to very high-energy gamma rays that have wavelengths smaller than an atom. Unlike other waves that need a medium such as sound, EM waves can propagate without a physical medium. Indeed EM energy propagates best in a total vacuum, as interactions with matter may impede or distort the transmission.

Light, and EM radiation in general, can be considered as both a wave and as a particle—a wave-particle duality. The photon is this elementary particle of light, a mass-less packet of energy. Some interactions and phenomena of light, such as diffusion, can be considered and easily explained by wave theory. Other aspects of light, such as many elements of laser physics, require a particle understanding of light. A comprehensive understanding of light (i.e., quantum mechanics) requires this dual wave and particle nature of light.

Related to, but distinct from, the weaponizaion of the EM spectrum is the particle beam, the use of atoms and the components of atoms, projected to a distance at speeds near to the speed of light, to provide a destructive effect. Though minute, the particles of a particle beam have mass. Like many of the technologies discussed in this chapter, low-power particle beams are in widespread use today. The cathode ray tube (CRT), found in older televisions and computer monitors, are built around an electronically aimed particle beam firing at a screen to produce light. Low-power particle beams are already used to sterilize food and medical equipment. At much higher-energy levels is natural ionizing radiation, which is in part made up of fast-moving atomic particles (the rest of it being high-energy EM radiation). The destructive effects of natural lightning, the passage of large quantities of charged particles through the air, provides inspiration for the concept, though not necessarily a means, of artificially generating these levels of energy, or providing the precision needed for a weapon. A weapon, unlike lightning, must be able to hit the same place as many times as commanded.

Particle beams were considered as missile defense weapons during early SDI research. Of particular interest was the neutral particle beam. The earth's magnetic field regularly deflects and traps high-energy particles from external sources, such as the sun, meaning that it would also deflect the beam of a charged particle weapon. A neutral particle beam produces a particle beam with no net charge, meaning it wouldn't be affected by the earth's magnetic field. A charged particle beam would also interfere with itself due to like charges repelling each other. A stream of charged particles would have a tendency to disperse due to being composed of elementary particles of one charge. This, however, adds the problem of how to neutralize the charge of a high-energy particle beam to the problems of generating such a beam in the first place. Later SDI research concentrated on kinetic-energy interceptors and the space-based laser (SBL) concepts as being nearer-term capabilities.

Lasers Part I: Basics

The development of the laser and related devices brought DEWs closer to plausibility. Light amplification by stimulated emission of radiation, or LASER, is an effect predicted by quantum physics in the early 20th century by Albert Einstein. The laser as a device became a demonstrated reality in 1960[1] and is generally credited to Dr. Theodore H. Maiman, although there is some dispute over this first laser, as well the place of earlier patents and scientific papers from the late 1950s concerning the specifics for a laser device before someone actual managed to build one.[2] The light component of the acronym would imply a device that produced EM radiation visible to the human eye, but the nomenclature has expanded to cover many devices that use the same science to produce other wavelengths of EM radiation. Based on the same physics, the microwave amplification by stimulated emission of radiation (MASER), basically a laser operating on the microwave portion of the spectrum was actually demonstrated earlier in 1954 by Dr. Charles Townes.[3] A MASER can also be called a microwave laser, and similarly a laser operating at the gamma-ray end of the EM spectrum, can be referred to as a GASER or GRASER.

The light that emerges from a laser is coherent. Critically for weapon applications, this means that all the photons of light are travelling in parallel. The unprecedented straightness of a laser beam has led to it being used as the very definition of direction in surveying, industry, science, as well as in defense applications. Measurements by lasers aimed at reflectors left behind by the U.S. *Apollo* moon missions are still used today for lunar research. Diffraction sets a fundamental limit on how far a particular wavelength of EM energy can be focused, leading to the more practical statement that a laser beam does not spread out very much over a useful distance, depending on how one defines useful distance. Shorter wavelengths of light will lose focus over longer distances. How tightly a laser can be focused, in other words the laser's practical range, is referred to as its "beam quality."[4] Not all laser technologies are suitable for weapons use due to inherent beam-quality issues. Specifically for discussions on laser weapons, the term high-energy laser (HEL) generally refers to a laser that combines both high-energy levels and an ability to focus such energies at militarily useful ranges.

In most lasers, coherent light is amplified in a substance referred to as the gain, or lasing, medium. Depending on the specifics of the laser, this gain medium may be a solid, a gas, or a liquid. Molecules of the gain medium are excited, that is, raised to a higher-quantum energy state. Specifically the electrons of the molecule are raised to a higher-energy level, with the particulars of the molecule defining higher-energy states. Energy states are discrete levels, there is no gradual increase or decrease, and an electron in an atom or molecule is always at a particular energy level, never in between the energy levels defined by the atom or molecule. When a photon of light interacts with an excited molecule of the gain medium the molecule returns to a lower-energy level, emitting the lost energy as a photon that is travelling in the same direction and phase as the first; the molecule has been stimulated into emitting a photon of light that is essentially of the same nature of the first photon. Of note is that the quantity of extra energy possessed by a molecule or atom in an excited state is the same quantity found in the emitted photon, which is the same as that possessed by the photon that triggered the release of energy. Therefore the wavelength, or type of light, produced by a laser is directly related to the particulars of the lasing medium.

Now in an actual device there are many millions of molecules that make up the gain medium, and these molecules must be in an excited state to be useful. The process for raising the molecules or atoms of the gain material into an excited state is called "pumping." Energy for pumping a lasing medium can come from direct electrical discharge, noncoherent light, another laser, chemical reactions, and nuclear reactions. Small semiconductor lasers, such as those found in DVD players, use milliwatts of electrical power. The nuclear bomb-pumped X-ray laser of SDI fame and controversy is envisioned as being able to direct a fraction of the power of nuclear explosion into a very powerful laser output before it is consumed by its power source.

At the point where there are more excited molecules in a gain medium than those in a grounded state, termed population inversion, stimulated emission results in a net amplification of light. Different materials have different limits on how much energy is required to reach this state and how much energy will actually smother the process. Most lasers are constructed for specific roles, with the energy source and gain medium determining output characteristics, including maximum power. The very first laser used a solid ruby crystal to generate a visible laser output. Successive generations of tiny semiconductor diode lasers produce first infrared, then visible red, and finally blue laser outputs needed for compact disk, DVD, and Blue-Ray technologies, respectively. Carbon dioxide and other gases are pumped by external power sources to produce lasers used for very short-range cutting and burning for medical and industrial applications. Chemical reactions generate a stream of excited molecules to be the lasing medium in many recent HEL weapon demonstrators.

With most lasers the gain medium is inside an optical resonator, with two mirrors or other means to reflect photons back and forth through the gain medium, stimulating the emissions of more similar photons, and building up strength until the light energy is released. In a mirror-based optical resonator one mirror is partially transparent to allow some of the generated light to emerge. The fiber laser is a major exception, and instead uses the same optical properties that allow a light signal to follow a winding fiber optic cable to bounce the photons repeatedly through a lasing medium in the core of the fiber. This configuration is of interest due to the flexible nature of fiber optics technology (relative to other optics), which allows greater freedom in a laser system's layout. Key to all HEL systems is high-quality optics that minimize absorption of the generated light, such absorption resulting in both loss of efficiency and damage to the laser itself.

The free electron laser (FEL) dispenses with a physical gain medium, instead using an electron particle beam that is manipulated by magnets into emitting a laser beam. There is no lasing medium per se as the electrons are subatomic particles. Magnets manipulating a high-energy electron beam cause the electrons to release energy in the form of photons. The nature of the photons released is defined by the input electron beam and the magnetic fields, resulting in a laser output. Through control of the magnets and the initial electron beam, the output laser can be varied in wavelength, meaning one device is capable of having its output "tuned" for specific applications and situations. The capacity for "tuning" the FEL output, as well as being powered purely by electricity, have resulted in this technology being described as the "holy grail" for laser weapons development. FEL technology is described as having "inherently high beam quality,"[5] meaning immediate potential for military ranges. For a long time, the U.S. Navy has been interested in an FEL-based defensive weapon due to the maritime environment causing great variability in the best wavelength for allowing laser energy to propagate.[6]

Since their invention, lasers have steadily found new applications. Among the most critical to modern life are lasers as tools of measurement, allowing for the precision necessary for many large-scale engineering and construction projects. These civilian-world applications are not too far removed from the military's use of lasers for weapons' guidance; the most precise munitions guidance systems are still based around seeking a dot of reflected laser light painted onto a target by a laser designator.[7] Like radar applications of the EM spectrum, lasers have proven useful as force multipliers. Advances since the Vietnam War have made such systems more reliable, lighter, and cheaper, but the principle remains the same.

A more sophisticated, but still broadly similar, use of laser technology is as the illumination source for light detection and ranging (LIDAR) and laser detection and ranging (LADAR) sensors. Beyond the simple LIDAR speed-measurement devices used to enforce speed limits domestically, are LIDAR and LADAR sensors capable of generating three-dimensional representations of objects. For robotics, such as the autonomous vehicles that competed in the three Defense Advanced Research Projects Agency (DARPA) Challenges,[8] LIDAR sensors were a common means to give these robots a view of the environment they were attempting to drive in. The capacity for LIDAR technology to obtain accurate data on the shape of objects is also useful as a means of target discrimination—a tank is of a different shape than say a minibus, and with high-enough resolution, different models of tanks may be discernable by LIDAR sensors.

Laser-guided weapons still rely on explosives and kinetic energy for a destructive mechanism. Though they have eye safety warnings, most lasers in operational use today are of relatively low power. Laser technology has yet to be employed, as casual futurists have been predicting, as a materially destructive weapon system.[9] However, the concept of an HEL weapon is an old one, and work on turning the laser into a weapon continues.

Lasers Part II: Stand-Ins and False Starts

Presently, gas dynamic laser technology represents the most advanced of HEL technology in terms of high-energy levels and beam quality needed for a weapon, but at the same time does not necessarily represent the best technology for a weapon. Gas dynamic lasers have demonstrated the capacity to generate multimegawatt energy levels. They have also demonstrated the ability to generate these high-energy outputs at wavelengths that facilitate effective energy delivery to tactically and even strategically useful distances. Finally, their effectiveness has been demonstrated outside of "clean-room" laboratory conditions. Two recent gas dynamic laser programs noted for their direct potential as fielded capabilities are the U.S. Airborne Laser (ABL) program and the joint U.S.-Israeli Tactical HEL (THEL). While providing quite a bit of research on HEL weapon technology, these programs also highlight many of the obstacles to using gas dynamic lasers as battlefield weapons.

The ABL uses a multimodule chemical oxygen iodine laser (COIL). Inside each COIL module, a chemical reaction and supersonic expansion nozzles generate energetic oxygen molecules, which in turn excite iodine molecules. These excited iodine molecules form the lasing medium and produce invisible infrared light. This high-velocity flow can be scaled up to produce megawatts of laser energy, while carrying away destructive waste heat. Several years of research spread over multiple ABL programs and a troubled development phase, culminating in the U.S. Missile Defense Agency's *YAL-1A Airborne Laser Test Bed* (ALTB), a converted 747 freighter containing six COIL modules, along with the optics train needed to aim the multimegawatt laser beam, including apparatus to detect and compensate for real-world atmospheric distortions, and onboard fire-control system. On February 11, 2010, the ALTB demonstrated its ability to destroy a ballistic missile while in flight (both ALTB and target missile).[10] Before the program was curtailed in 2009 to one research aircraft[11] there was even some discussion on a small fleet of operational versions of the *YAL-1A*. Aside from severe fiscal constraints in 2009 and subsequent years, and the *YAL-1A* program being several years behind schedule, Defense Secretary Robert Gates noted other "significant"[12] technical issues with the concept in his April 6, 2009, presentation on the 2010 defense budget. Among the criticisms leveled against the idea of directly turning the *YAL-1A* into an operational warplane is the low range of the so-far developed HEL system, thought to be only several hundreds of kilometers.

The common mission envisioned for the ABL program in its many guises over the years, is the boost phase destruction of ballistic missile targets. A ballistic missiles flight, for the purposes of missile defense, is divided into three phases or components: boost, midcourse, and terminal. During the boost phase, the missile's propulsion is firing, which importantly for missile defense means: (1) the missile is still accelerating to the velocity needed for a ballistic trajectory to deliver warheads and (2) the rocket propulsion provides an easily detected and tracked heat signature. Politically, the missile also may have the benefit of not being over the target nation yet; with less-sophisticated threats, the missile is still over the offending nation, meaning that missile destruction minimizes endangering the "innocent."

A laser weapon is applicable to boost phase missile defense on many fronts. First, the beam is travelling at the speed of light; therefore, as long as the beam connects, the attack is practically instant at the ranges being considered. This then leads to the possibility of rapidly shifting to other targets, or re-attacking if the first attack was insufficient. Now for a laser weapon to be an effective boost phase defense, its range and ability to deliver energy must be considered. Unlike a physical interceptor missile, such as those used by the U.S. Navy, which takes time to reach the target, but have a reasonably immediate effect on striking the target, the destructive effects of these initial HEL weapons require some exposure, or "dwell," time. In the ABL, the "kill mechanism" is simply the laser heating a small patch of the missile's exterior to the point where the skin or the structure fails. Rocketry in general is noted for its relative fragility, and anything that takes a missile out of its operating envelope, such as a weakened structure or severely distorted skin, would result in its destruction. Liquid-fuelled rocketry is considered to be more fragile than its solid-fuelled counterparts; however, the relative sturdiness of solid-fuelled targets can be accounted for by increasing power margins, or the time the laser dwells on the target. The subject of lethality highlights the fact that simply achieving a multimegawatt power levels for a laser is insufficient for it to be directly employed as a destructive weapon.

A megawatt-class laser power only brings the potential for target destruction. To be an effective weapon, the laser must be able to deliver enough energy to achieve a destructive effect quickly enough to justify its use over a physical attack. Given a high-enough rate of energy delivery, rapid heating of solid metal becomes rapid conversion to vapor or even superheated plasma—in effect an explosion; however, this is beyond the energy levels produced by the six COIL modules used in the ATBL. A limited energy delivery rate also complicates the targeting problem. The longer the laser must dwell, the longer the target must be precisely tracked to maintain the rate of energy delivery. This is not just a problem of getting energy to the target missile, but to a small patch on the missile. Warming a large portion of the missile does not lead to the same rapid destructive effect as melting through a point on the target.

Nature also may impose a limit on how much energy may be delivered. Some laser energy is absorbed by the atmosphere as the beam makes its way toward the target. In fiction, a laser is portrayed as a visible beam reaching out from the weapon to the target. In reality, effects visible to any observer not directly along the axis of the beam are wasted energy. Dust and other airborne particles may scatter laser energy away from the target and in the process produce an eyesight hazard. Like the target, as well as the laser system itself, particles in the air, rain drops, and the gaseous air molecules themselves can absorb laser energy, heating up in the process. Besides reducing the power delivered to the target, atmospheric effects such as thermal blooming complicate the already difficult feat of keeping a beam focused on a moving point at ranges measured in hundreds of kilometers. Air at different temperatures has different optical properties, such as can be seen in the shimmering above a hot surface. The higher the energy is, the worse the effects are. Essentially the passage of high energies through the atmosphere provides a source of atmospheric distortions that inhibit the passage of high energy through the atmosphere.

The ABL program is known to combat the problems of high-energy delivery within the atmosphere in two ways. First the iodine lasing medium of the COIL system produces light at a wavelength (1.315 microns) that minimized absorption by the atmosphere.[13] To counteract natural and remaining self-generating atmospheric distortions, a system of adaptive optics predistorts the beam so that the attack laser arrives focused at the target. Among the components of this active compensation system is a low-power laser sensor that measures the amount of distortion present. High-speed computer analysis of the external distortions generates commands to predistort the beam via mechanisms in the optics train before the beam leaves the aircraft.

Though not directly part of the ABL program, the advance tactical laser (ATL) also uses COIL technology—this time a small installation onboard a C-130 Hercules transport aircraft. This system is only meant to generate hundreds of kilowatts of laser energy, and has a range measured in dozens of kilometers. Unlike the ABL/ATBL, the ATL program is meant to study the use of a laser against ground-based targets; the Special Forces community is interested in its potential. According to the December 2007 Defense Science Task Force on Directed Energy Weapons, ATL is at an earlier stage of development than antiballistic missile defense oriented ABL work.

Using a different combination of chemicals are the ground-based THEL and related mobile THEL (MTHEL) programs. Although THEL is a lower-power system, with a shorter range, and ground basing that is technically less demanding, it has been able to destroy a wider variety of threats, including thick-skinned mortar shells. Israeli participation in the THEL program is related to the ongoing threat from short-range rockets, artillery, and mortars employed by terrorists. Low-tech rockets, artillery, and mortars, labeled RAM, and are an increasing threat due to their low cost, portability, and easy concealment in urban environments. Once fired, RAM threats provide little time for evacuation or defensive actions, often leaving retaliation and, more controversially, preemptive attacks in neighboring territory as the only options. The very low per-shot cost of RAM threats presents very serious economic problems to defense, where million-dollar interceptor missiles are used to counter rockets priced only at hundreds of dollars. Claims both for and against[14] an operational THEL put the per-shot cost for the chemicals required at thousands of dollars.

Like ABL, the THEL program generated some discussion on a deployable version, including the MTHEL. Also like the ABL program, THEL fell behind schedule, again due to the many technical challenges associated with attempting to weaponize laser technology as a kill mechanism. THEL successes include destruction of in-flight short-range rockets, mortars, and artillery shells fired in salvo.[15] The THEL program was shelved in 2005, a move decried by some politicians in Israel as the RAM threat still exists. On the other hand the cancellation revealed several troubling problems with the THEL concept at that point in time.

One inescapable issue with gas dynamic lasers is that their "fuels" tend to be bulky and often hazardous as well. Environmental concerns may limit the selection of prospective reactants, contributing to factors that have stalled enthusiasm for operational use.[16] In turn, many current chemical laser programs and concepts counter such concerns with the option of an exhaust scrubber, and containment of exhaust in a sealed system.[17] The amount of chemicals needed to operate these gas dynamic laser-based weapons also adds to the bulk, leading to deployment and vulnerability concerns. Earlier more ambitious laser-research plans of the United States included on-site and in-flight reprocessing of laser reactants, allowing for potentially limitless magazines, as long as power and time were available.[18] A regeneration system for recovering the chemicals needed for the laser would add, however, to the bulk and complexity of what is already a difficult-to-put-together system of systems.

The reagents used in the COIL are gaseous chlorine, molecular iodine, and an aqueous mixture of hydrogen peroxide and potassium hydroxide.[19] Although these chemicals do seem relatively commonplace, with forms of each being found in bathroom medicine cabinets and in cleaning supplies, in the concentrations and purity needed for COIL some are quite dangerous. Chlorine gas has a reputation as a poison used in chemical warfare, and even lower concentrations used in water purification require protective equipment. Competing chemical mixes fare no better. The deuterium fluoride (DF) gas dynamic laser that THEL is based on is fuelled by precursor chemicals that are described as toxic.[20] Though exotic sounding, deuterium is simply a stable isotope of hydrogen, and can be extracted from natural water sources. The reactants that supply the fluorine on the other hand are described as both toxic and corrosive.

It must be remembered that laser technology is competing with physical-interception methods, such as interceptor missiles. Moore's law implies that the foundation technology for the sensors and guidance, electronics, becomes more powerful exponentially.[21] Guided missile technology is a familiar technology, and despite many initial problems, especially with reliability, has matured over the years and the basic technology is generally trusted today. An example of this competition is the Israeli *Iron Dome* counter-RAM interceptor missile system—a system procured after the end of the THEL program. Though missiles do cost more on a per-shot basis, there are not quite as many risks associated with producing an operational system. *Iron Dome* was fielded in 2011 and has already claimed some operational success during rocket attacks in early 2012.[22]

Chemically powered lasers, having demonstrated the ability to reliably generate HEL outputs, continue to prove useful in the development of ancillary systems, such as the optics train needed to accurately direct the beam. This is the purpose now of the *YAL-1 ABL* research on the technologies needed to field an HEL weapon. However, experience gained from chemically powered laser work has highlighted several problems that make operational deployment of this specific method of beam generation unlikely.

Lasers Part III: Electric Lasers and the Promise of Limitless Defense

Since the prominent ABL and THEL/MTEL gas dynamic chemical laser programs were curtailed, U.S. laser research has continued with emphasis on weaponizing electrically powered laser technologies. An electrically powered laser avoids many of the logistical problems of supporting a chemically powered laser, not the least of which is lacking the need to supply large quantities of exotic and potentially hazardous chemicals. The lack of an additional vulnerable logistical trail is just one factor in the perceived robustness inherent to many electrically powered laser concepts.

Although the science backs up the potential for many gain mediums to produce megawatt power laser outputs, engineering challenges have so far kept power levels at kilowatt levels. One immediate problem is dealing with the power levels involved, and the generated waste heat. Inefficiencies in conversion of energy, electricity to light, result in the generation of waste heat. In 2010, companies were making bold claims of better than 30 percent efficiency,[23] meaning the destructive energy delivered by such a laser would be only around a third of the energy handled within the laser system itself. Now of course these energies would not be concentrated on a tiny point inside the laser, but a slow cook can be just as destructive as the weaponized effects on the target. As many of these current laser proposals are for close-in weapons systems (CIWS), the final line of defense against salvos of guided missile and lately against multiple inbound RAM threats, the lasing material may not have much time to cool down between shots. It is not surprising that among the technologies promising to be the key to a working defensive laser weapon are high-temperature ceramics.

Although benefiting from the lack of a physical lasing medium that may melt or boil away, the FEL must contend with being able to generate a high-energy electron beam capable of in turn generating an HEL output. At the time of writing, the U.S. Navy announced a major breakthrough in this area, leading toward megawatt power levels down the road for its FEL program.[24] Though the existing 14-kilowatt prototype does not have the power needed for a weapon, the navy's demonstration of high power level electron beam technology is notable in the troubled history of laser weaponization for being ahead of schedule.[25] Despite this hopeful news, it must also be remembered that even optimists estimate that at-sea testing for a 100-kilowatt system would be feasible around the year 2020,[26] and this is contingent on funding. It would still be some time after this initial at-sea demonstration that the megawatt system hinted in the press release could become feasible. Again, the problem is dealing with the input energy requirements needed to function.

Lately, about the only part of the energy problem for electrical lasers where a solution is readily available is power generation, though with caveats.

Naval DEW mounts would only be additional items on the list of electricity-hungry applications, such as advance radars, instead of a special requirement for power generation. If a warship is able to generate onboard the electrical power needed to power a DEW then a fixed ground installation would be easily conceivable. Generating and storing the power needed for an airborne electrically powered laser is another area of concern; however, the gas turbines powering many of today's warships often trace their lineage back to commercial airliner power plants. Smaller mobile power sources are, however, still a problem; however, the interest in "green" technologies such as, again, electric transmission, but this time for cars and trucks, is presenting a range of energy options for vehicle-mounted direct energy weapons. An electrically powered SBL would certainly benefit from advances in "green" power, such as increased efficiency solar panels and lightweight power storage.

The low power levels achieved thus far by electrical laser technology has not stopped these technologies from reaching the battlefield. Highlighting recent progress on electrically powered solid state lasers is the battlefield deployment of ZEUS-HLONS (High-Mobility Multipurpose Wheeled Vehicle [HMMWV] laser ordnance neutralization system), a relatively off-the-shelf industrial laser mounted on an HMMWV light vehicle. ZEUS-HLONS is meant to burn through and ignite IEDs and land mines for a slow burn as opposed to setting off a detonation. Explosives used by militaries and industry generally only detonate under specific circumstances, often burning before they would detonate. Standoff range is useful to such roles as not all bombs are composed of such safe explosives, and an ignition may still result in a lower-order detonation. This program has seen use in Afghanistan and Iraq, two conflict zones where IEDs are a severe threat to U.S. and allied forces. Unlike in-flight rocket, artillery shell, and mortar threats, taking minutes to burn off IEDs and unexploded ordnance represents a viable strategy and improvement over the time needed for manual ordnance disposal. This and other concepts for using off-the-shelf lasers for niche applications are useful in building acceptance of lasers on the battlefield after so many years of confinement to laboratories and being laughed off as fiction.

The kilowatt-class laser being used as a bomb disposal tool also represents how the concept of the laser as a weapon has evolved with the threats of the day. In the 1980s, the Cold War was the preeminent security concern for the United States; HEL technology was shortlisted in the SDI program as a possible defense against Soviet ICBM attack. In the 1990s, the tactical ballistic missile threat posed by rogue states, such as Iraq under Saddam Hussein, drove HEL development. The ABL program originally started off in this period with the aim of countering the more localized threat posed by tactical ballistic missiles. Since 2001 and the start of the Global War on Terror, a major role being promoted for HEL development is in countering asymmetrical threats posed by subnational entities, such as IEDs and unguided RAM threats.

Although HEL weapons may be used offensively, they have over the decades been promoted mostly for defensive roles, and specifically for providing a shield against inbound munitions. Part of this is of course avoiding negative perceptions associated with developing what is basically an entirely new and untried kill mechanism. This is not just avoidance of criticism from pacifist communities, but also from existing weapons communities whose budgets and projects may be threatened by this futuristic technology. Of the recent U.S. DEW programs, only the airborne tactical laser is openly being promoted as only a research program to investigate the utility of a destructive laser weapon in support of forces on the ground,[27] offering both precision and novel effects to set it apart from existing capabilities.

Defensive weapons, specifically weapons that attack munitions mid-attack, on the other hand represent a relatively new capability. Existing antimissile capabilities are often likened to "hitting a bullet with a bullet," in past a charge by critics to indicate futility at the effort, but now more of a boast from counter-missile system supporters as the concept has at least been made viable by modern computer technology. In this sense, lasers and other DEW-based antimissile systems would seem to be redundant; however, the same technology that has allowed missile defense interceptor missiles to be successful is also applicable to threat missiles. Also, new categories of threat have emerged from technology proliferation. Instead of attack by a single highly capable missile, low-cost electronics allow the construction of swarms of low cost, but still capable, munitions to overwhelm a defender. The asymmetrical threat environment now emerging presents several challenges that can be met by DEWs due to their potential for speed of attack, and potential for "limitless magazines."

EM radiation propagates at the speed of light, 299,792,458 meters per second,[28] meaning an EM-energy-based weapon will hit its target for all practical purposes instantaneously. In the face of both stealthier and faster (supersonic and hypersonic) missile threats, there is a fear that physical missile interceptors will not have the capacity to maneuver to successfully intercept in the shrinking time window between detection and impact. Vertical launch systems have become something of a norm for interceptor missiles, in part due to their capacity to rapidly launch against multiple targets compared to reloadable missile launchers, the missiles being responsible for actually pointing itself toward the target once clearing the launch tube. The aforementioned *Iron Dome* system uses a deployable battery of vertical launch tubes. Vertical launch works in part because the interceptor has time and propulsion capacity to rapidly point itself toward the target before actually accelerating toward the target. This flight path from launcher to target involves a rather dramatic turn: the greater the rate at which a missile has to change direction, the greater the G-force involved. Future threats may not present enough time for such a maneuver, or physical technology may not be robust enough to conduct the necessary form of maneuver.

For a DEW-based defense to be competitive, it must be capable of rapidly and reliably producing a destructive effect. One DEW "turret" can only attack a salvo threat one target at a time, whereas many missile systems can conduct multiple simultaneous interceptions. Although this comparison seems to be to the detriment of DEW, the impression breaks down in the details: the number of multiple simultaneous intercepts is limited by the missile system's capacity to have multiple interceptors in flight at any one time. Fire-and-forget technologies do mitigate the fire-control burden, but increase the cost of the interceptors. Also, the time it takes for an interceptor missile to conduct an attack, and the time for a miss to be recognized, may not leave enough time for another interception to be attempted. Generally multiple interceptors would be launched simultaneously against one target to increase the probability of an intercept; this, however, has consequences again on the number of targets that can be attacked. If a DEW-based defense can destroy more targets during the flight time of a missile-based system attacking its theoretical maximum number of simultaneous targets, then the DEW-based system is superior in the rate of interception due to not needing multiple interceptors. Therefore, scaling up the power of the laser to where it can outperform legacy technologies becomes the primary technological barrier, as without a significant increase to how fast individual threats are destroyed, a DEW would be no better than physical interception systems.

One limitation that the missile battery and the chemical-powered lasers of the THEL and ABL programs share is that they only have a finite number of shots. The electrically powered laser on the other hand will fire as long as power is supplied to it, and as long as it remains intact. Electrical power is reasonably plentiful, but much research and development is needed to first generate a worthwhile laser output, and then having that laser operate reliably. Once these technological challenges are overcome, then an electrically powered laser weapon would have a practically unlimited magazine. A "limitless magazine," plus a faster engagement rate, may not equate to an impenetrable shield, but it may be necessary to keep up with saturation threats.

As mentioned earlier, one potential metric of the effectiveness of a defense is the per-shot cost relative to the cost per shot of the attack (this measure, however, is often ignored due to the overriding value of the asset being protected); intuitively, an electrically powered laser defense could excel in per-shot cost effectiveness. Ignoring for the moment the cost of what is still expected to be a long research and development period, we could say that electrical power is plentiful in many situations right now and is expected to only become more plentiful due to other demands for generating capacity. Keeping a laser- or other DEW-based defense operating would be a matter of burning more fuel to run the generators.[29] Logistically, this would give an electrically powered direct energy weapon an advantage as necessary fuel and ammunition are one in the same.

Although the chemically powered laser is a technology that is much nearer to being a fielded capability in that it has demonstrated the power levels close to that needed for a weapon, it is not the form of laser technology that offers the full promise generally advocated by DEW supporters. Aside from the many problems experienced by the specific ABL and THEL programs, these two systems could also be considered interim capabilities, potentially starving funding from both the R&D needed to mature higher-payoff electrically powered systems and from refined methods of physical attack that are still effective in the short term. This also highlights again that the timeline for a deployed destructive laser weapon, outside of a few niche applications such as bomb disposal, will still likely be years, if not decades.

Indiscriminate Energy Weapons

In a way energy weapons have been available since World War II through the EM effects of a nuclear detonation. Depending on where the detonation takes place relative to the surface of the earth, different mechanisms for damage will be enhanced. A surface or underground detonation, expends some of the device's energy against the ground, limiting blast effects. A correctly situated airburst will magnify the blast effects as the shock wave of the explosion is reflected back on itself by the ground (this works for conventional explosives as well). At a high-enough altitude, the energy released by a nuclear detonation with interact with molecules in the upper layers of the earth's atmosphere, as well as the earth's magnetic field, to produce wide-ranging EM effects. Basically this final configuration for detonation turns the nuclear weapon into an indiscriminate energy weapon.

High-altitude nuclear events or explosions (HANEs), also known as high-altitude nuclear detonations (HANDs), have many destructive effects on computers and electronics. A portion of the energy released from a nuclear detonation is in the form of high-energy particles. These high-energy particles can get caught in the earth's magnetic field, just as natural high-energy particles do. In both cases these temporarily trapped particles form radiation belts, the Van Allen belts being the natural ones. When there is an overflow of particles, beyond enlarging and enhancing the natural Van Allen belts, extra radiation belts form in regions that usually do not have them. For satellites not designed for prolonged operation in the natural-radiation belts, repeated exposure to these unexpected zones of highly energetic particles will drastically cut down on component life and lead to very early satellite failure.[30] These artificial belts naturally lose particles over time, as they are not continuously replenished by sources such as the sun, but can take years to dissipate.

On earth, the energy released by the HANE will setup various line-of-sight and non-line-of-sight EM effects. This includes high-energy particles bombarding the sparse molecules of the upper atmosphere, the energy of which is then reradiated as EM energy. The EM energy unleashed by a nuclear detonation interacting with the earth's magnetic field results in geomagnetic effects. The high-energy EM pulse (EMP) of the detonation itself, and reradiated EM effects in the radio-frequency (RF) portion of the spectrum, can interact with electronics and electrical equipment, setting up unwanted signals and electrical currents that may burn out both large circuit breakers and the microscopic logic gates of modern electronics. Electrical infrastructure is vulnerable from the long transmission wires; EM interactions will setup currents and high voltages that may destroy critical electrical transformers. Scenarios have been given where a single, moderately sized, nuclear detonation is able to wipe out the electrical infrastructure of the entire continental United States.[31]

The damaging effects of a HANE are not theoretical. Solar weather can also result in similar effects as a HANE, including overwhelming the natural outflow of particles in Van Allen belts, which result in satellite damage. Severe solar activity can make the northern and southern lights visible at lower latitudes. The EM effects of natural solar activity has on occasion knocked out entire power grids, as it occurred in the province of Quebec, in Canada, during particularly active solar activity in 1989. These natural events both add to the knowledge base concerning EM weapons and serve as additional incentives to harden infrastructure—EM warfare is not a certainty, but severe solar weather, like earthquakes, are only a matter of time.

For the United States the most well-known demonstration of the damage a HANE could generate was by accident in the consequences of the 1962 Starfish Prime nuclear tests. The tests had a nuclear detonation of around 1.4 megaton yield, at a mere 400 kilometers altitude. This detonation had several unexpected results, including the generation of extra radiation belts that damaged several satellites over time and a larger-than-predicted EMP. This EMP is credited with knocking out traffic lights and causing other EM interference in Hawaii (roughly 1,500 kilometers away from the Johnson Island test site).

It must be remembered that the power, telecommunications, and other electronics that were disrupted and in some cases damaged, were from the 1960s and are generally regarded as being somewhat more robust than today's computer devices. Modern computer circuits are minuscule in comparison, and operate on a fraction of the power. Static electricity from a careless technician is enough to destroy a modern computer chip. Also, it must be remembered that in the 1960s practically no aspect of day-to-day life was reliant on computers, whereas today a strong argument is made that Western civilization is computer dependent. Indeed, the United States of the 1960s was not quite as dependent on electricity as it is today. Therefore, although nuclear weapons and delivery technology has become more potent, Western society has become more vulnerable outside of expensively shielded and hardened military equipment.

Although none of these effects, except perhaps the blinding light of a nuclear detonation, are directly harmful to people, a wide-area EMP attack perpetrated against the United States would in the long term lead to significant loss of life. The disruption to medical services due to loss of power, communication, and potentially direct failure of life-support equipment, would be an indirect casualty mechanism causing significant loss of life within minutes to days. The loss of power to utilities would lead to water shortages in many locations, leading to further casualties over a period of days and weeks. Disruption to food production and transportation could eventually lead to mass starvation in only a few months. The disruption to fuel supply and transportation would also leave many trapped in the dead cities of this worst-case scenario. Globally there would be economic repercussions as the largest economy is suddenly removed from the grid. Simultaneous and permanent failure to many critical elements of the North American power grid could take years to repair and would be a cripplingly expensive recovery.

A HANE attack would of course have military repercussions. In orbit, satellites would begin to fail, including many critical to the military space-force enhancement, limiting options for a conventional response. This would be true if the detonation was situated to avoid damaging terrestrial infrastructure. The loss of a conventional military option, at least one supported by space systems as has been the preference since the 1991 Gulf War, raises the problem of what to do next. The HANE attack being perpetrated by another nuclear power gets mixed in with nuclear deterrence and the controversy over whether a limited nuclear attack between nuclear-armed powers was possible without escalation. According to the 2004 *Commission to Assess the Threat to the United States from Electromagnetic Pulse (EMP) Attack*, post–Cold War Russia and China have considered the idea of using the EMP effects from a limited nuclear strike to "paralyze" the United States.[32]

The U.S. nuclear forces are meant to be hardened against the effects of a general nuclear war, and despite the end of the Cold War, and the de-emphasis of the nuclear mission by both the Bush[33] and Obama administrations, the United States still maintains all three elements of the nuclear triad of bombers, ICBMs, and a fleet of ballistic missile submarines. Tactical nuclear war, or the limited use of nuclear weapons, is problematic, in that it may escalate to all-out nuclear war. All-out nuclear war leads to the still-applicable Cold War concept of mutually assured destruction (MAD), wherein enough retaliatory forces exist such that one power could not ensure the loss of the other's ability to deliver a crippling retaliatory attack.

Proliferation, however, has meant that the technology needed for a HANE attack has spread to others who may not fit into the model of nuclear-armed great powers deterring each other with reasonable effectiveness. The concept of rogue states, nations led by leaders who regularly flaunt international norms, includes the possibility that nuclear deterrence may not work against these regimes. Indeed there are fears that proliferation may eventually deliver nuclear weapons into the hands of terrorists, subnational entities that exist outside of the international system of nation states. Put simply, nuclear deterrence would not work against terrorists in that they would not have anything worth using a nuclear attack to retaliate against.

One possible scenario is the nuclear antisatellite (ASAT) employed by a newly armed nuclear state, wherein a HAND would not be able to disrupt life directly in the United States, but instead destroy the satellite infrastructure needed to intervene militarily in a regional conflict. This scenario is applicable to an emerging nuclear-armed nation, one that does not have the ability to threaten the United States itself, but simply wishes to prevent the United States from acting in its vicinity. The loss of military space-force-enhancement assets would hamstring conventional forces. If the disruption was great enough, the only way to proceed with conventional warfare would be to accept high casualties and collateral damage. Nuclear retaliation would bring its own moral and ethical problems. A newly armed rogue nuclear state would have few, if any, nuclear weapons remaining after its nuclear ASAT bid to prevent U.S. intervention. Moreover, it has limited its use of nuclear weapons to only prevent the deployment of global U.S. power. Would the loss of critical, but largely unmanned, infrastructure to intervene in a foreign conflict warrant a nuclear response? Related to this would be the loss of commercial and other nonmilitary space infrastructure, much of it necessary for the modern world to exist. Would destroying a foundation of the Western way of life, but again without massive direct loss of life or an immediate threat to the homeland, warrant nuclear retaliation? Is proportional response sacrosanct, or would this warrant a nuclear strike to deter future rogue states? This troubling scenario the author has in previous works referred to as a possible way to "effectively get away with using a tactical nuclear weapon,"[34] or in other terms, the nuclear ASAT is the tactical nuclear weapon of the information age.

High-Power Microwave (HPM) and Radio Frequency

The preferred method of warfare in the West, and the United States in particular, is one of precision. Aside from the laser, there is interest in direct energy weapons that attack electronics through RF, but not of the indiscriminate nature presented by a HANE. Some of this is in dedicated programs to build RF weapons, similar to the concept of the radio "death ray" noted at the beginning of this chapter. Other capabilities are falling out of the high transmission power levels inherent in many radar-sensor suites. In light of the fact that some early radar research had roots in trying to create a DEW, it is somewhat ironic that radar sensing, RF jamming, and finally RF attack are converging in some multipurpose radar arrays being marketed today.

Historically, radar and radio transmitters have been known to unintentionally disrupt aircraft systems. Shades of this are seen on every commercial flight as the cabin crew warn passengers to turn off mobile telephones and other electronics. All aircraft are shielded to some degree from EM interference (EMI); however, in operational use, shielding may become damaged, or may encounter a signal source it was not meant to protect from. Conversely emission standards for electronic equipment are meant to prevent internal generation of damaging signals; however, again devices may on their own malfunction in this regard. Among the many difficulties and costs of systems integration are getting fully functional electronic devices to work together inside cramped avionics and electronics bays. Finally, there are existing electronic warfare (EW) systems, some of which can radiate enough energy to damage radar components.[35]

Radar, radar warning, and communications antenna are naturally vulnerable points with respect to EMI as these are intentional and necessary openings for EM energy to pass through. Radar and communication jamming simply overwhelm a particular system with RF noise, degrading its ability to receive anything intelligible. For electrical engineers on military projects, the problem is how to have these openings without compromising the EMI resistance overall. Again, all this contributes to the high cost of weapons platforms today. Shielding itself can be overcome with enough energy.

The world of EW has, since its beginnings, been shrouded in secrecy. For many, EW is an arcane and unseen battle between black boxes. Although the results are quite important, they are generally invisible, adding an additional layer of obscurity. On the other hand, the use of RF and microwaves in a destructive mechanism against more than signal reception has been investigated since at least the 1930s, with no tangible results, leading to a degree of skepticism over this line of research leading anywhere. A laser weapon is something the general public and policy makers at least have some concept of from fiction. EW is both intentionally hidden away and unintentionally misunderstood by the public at large and possibly by many policy makers.

In the very near term, a single-shot EM weapon is feasible. Since the mid-1990s, Australian defense analyst and electrical engineer Dr. Carlo Kopp has been discussing the potential to build such a weapon involving an explosively driven power system coupled to the proper transmitter.[36] Although not commonplace, there is nothing that is part of science fiction about converting a fraction of the mechanical energy released by high explosives into electrical energy. A high-explosive detonation has a timeline, by definition a burn rate higher than its speed of sound, which means plenty of time for this energy to be harnessed via electromechanical apparatus. Indeed there are a range of technologies, including the explosively pumped flux-compression generator, described by Dr. Kopp as a "mature technology" in the 1990s, and magnetohydrodynamics (MHD) technologies.[37] The latter has also been applied in less-destructive forms of electrical-power generations, having being used to extract additional electricity from waste heat in existing thermal power stations. Here again the technology may sound fictitious, but it already exists and is in use internationally.

Although perfectly fine for an EM warhead, the idea of using cartridges filled with explosives and electronics, like the chemical-supply reality of existing HEL demonstrators, does not fulfill the "limitless magazine" promise of DEWs. Again, for many installations the power supply of a warship is considered necessary for initial deployments of RF and HPM-beam weapons. Naval interest in RF and HPM is due to this type of EM energy's potential to cut through atmospheric conditions, such as fog, found during maritime operations. Indeed real and persistent criticism of laser technology is the ease at which it may be blocked by dust and other environmental factors, let alone intentional attempts at disrupting the conditions needed for beam propagation.

RF and HPM weapons are beginning to emerge into the public eye. In 2008, the U.S. air force tendered for equipment to test the feasibility of airborne HPMs that would be able to interfere with, and possibly destroy, targeted electronics.[38] This was all done in the public, as opposed to some very secret "black" program such as early stealth work. Aside from being a multishot system, and having a very modest budget of about $40 million,[39] the counter-electronics HPM advanced missile project (CHAMP) is aiming at being flown on an unmanned aerial vehicle (UAV). Boeing was awarded the contract in 2009, with a press release noting that this will be the first time an HPM weapon has been integrated with aircraft.[40]

In perhaps another sign that RF and HPM may be about to emerge as a viable defense technology, Raytheon is marketing the *Vigilant Eagle* system for protecting airliners from the threat of man-portable air-defense (MAN-PAD) missile systems. Terrorists have already demonstrated both the ease at which MANPAD systems may be acquired, and an ability to bring them within range of civilian runways.[41] Basically *Vigilant Eagle* is a ground-based system for rapidly detecting, and attacking with EM energy, portable and easily smuggled antiaircraft missiles. The EM energy projected by *Vigilant Eagle* is meant to interfere with the electronics of a MANPAD and other short-range air-defense (SHORAD) missile. Precision over the beam and its effects would be absolutely essential for this system to avoid itself becoming a risk to civil aviation.

These two examples of recent EM weapons work are short-range systems; CHAMP is UAV mounted to get it near targets without risking an aircrew, and *Vigilant Eagle* is meant to provide local defense against very short-range missiles. In general, RF and microwave beams do spread out more than lasers do, though not as much as incoherent light (in or near the visible portion of the spectrum). Radio waves and microwaves are on the larger end of the EM scale; therefore, these technologies have fundamental limits on how far they can be focused. That being said, RF and microwaves are largely immune to being blocked by atmospheric conditions (for instance radar can "see" through fog, while your eyes working on visible light cannot). Also RF and microwaves can penetrate through materials that are mostly opaque to other parts of the EM spectrum. Indeed that is the point of an RF or microwave weapon—to penetrate not just into a target, but into the electronics that make it a high-technology threat. Finally, being relatively shorter range compared to the theoretical focus limit of a laser (which in atmosphere systems still have far to go before reaching) is still a militarily useful distance at several hundreds of meters.[42]

This emphasis on targeting the "black boxes" of a target does, however, raise the problem of battle-damage assessment. Unlike an HEL, which is meant to provide a visible effect of an exploding missile or shell, destroying the ability of electronics to function correctly does not leave much in the way of immediate signs. A successfully attacked computer system, to most conceivable methods of remote observation, looks identical to a computer system that is simply turned off. Part of EW is detecting that one has been attacked, which opens up the opportunity to simply "play possum" and turn off the attacked, but still working, system in anticipation of surprising the attacker. For a system such as *Vigilant Eagle* there is the additional problem that a successfully attacked missile may still impact the airliner being protected. Unless the electronic interference causes the threat missile to violently go off course or prematurely detonate there is no way to know if the threat was neutralized. Indeed, if the missile's electronics are simply burned out, the missile may still be flying toward the unaware airliner with a chance of hitting it.

The problem of battle-damage assessment is linked to the question of where electronic attack fits on the battlefield. Is it a means of support—to increase the odds for a traditional weapons' platform to attack with traditional "kinetic"[43] munitions? Or are the energy levels involved in an RF and HPM weapon sufficient that these systems can be used directly as the only means of attack? In an era of tight budgets, the next natural question is how much can be spent investigating the promise of RF and HPM weapons? Which is of course countered by the question: what is the price for not increasing the budget to include RF and HPM weapons?

The other side of RF and HPM weapons research is the vulnerability of the U.S. military and Western society in general to such electronic attacks. A benchmark for today's hi-tech society is the sheer proliferation of computer chips and wireless networking; both elements are connected to an earlier foundation of modernity, the electrification of society. As mentioned earlier, both Russia and the People's Republic of China have in recent doctrines considered the option of using nuclear weapons to achieve the effect of shutting down modern life in the United States. For near peers, the smaller area of effect from EMP warheads based on nonnuclear technology would avoid many of the problems of the "limited nuclear war" concept. Also by being nonnuclear, such systems would face few barriers to open proliferation in the world's arms markets. Nonconventional-warfare technology, such as GPS jamming equipment, meant to nullify the bulk of U.S. guided weapons is available for sale, and has even seen combat use against the United States.[44]

Less-than-Lethal Weapons and the Potential for DEW to Save Lives

On one end of DEW research is the problem of delivering sufficient energy to rapidly destroy, whereas on the other end the problem is delivering limited amounts of energy to achieve coercive effects but otherwise remaining harmless to the target. In general these systems face controversy not over feasibility, but over the potential for misuse. Directed energy, used in a less-than-lethal manner, is a new capability, with many unknowns as far as doctrine, capabilities, and risks are concerned. Limited understanding of these systems has led to many misconceptions about the technology.

The term "less-than-lethal" replaces an earlier term "nonlethal," when describing the same thing. Although a seemingly more friendly term, "nonlethal" gave rise to the incorrect impression that systems under this label had no lethal potential. Used incorrectly, many less-than-lethal technologies do have the potential to cause loss of life—rubber bullets and batons (clubs) can and do kill. Less direct, but just as deadly, would be the case of mass panic triggered by the use of a less-than-lethal weapon. In such mass-panic incidents there would be great potential for trampling and asphyxiation as direct casualty mechanisms. Of course lethal weapons could also lead to the same circumstances, which brings up the question of whether a technology can be made culpable when assigning blame in the aftermath, and if blame is assignable whether it was more a problem with the doctrine, and those implementing it, rather than the equipment.

Lack of understanding over the capabilities and limitations of a tool contributes to it being used incorrectly. Amnesty International, in a 2008 report on the use of "stun weapons" in the United States affirmed the organization's support for the development of less-than-lethal weapons technologies, as encouraged by international standards,[45] but was very concerned over the potential for misuse, and the safety of such devices. Among the issues raised was that the use of "conducted energy devices" (CEDs), or more commonly "Tasers," named after the most well-known brand, that have gone beyond being an alternative to lethal force, and had a lower threshold for deployment. It should be noted that the scope of the report was concerning the use of CEDs by law enforcement in the United States, a nation with significant oversight and limits on law enforcement, as well as other avenues for legal recourse for those who feel they have been wronged by the authorities.

Criticism that less-than-lethal weapons technology has a greater potential for "overuse" is related to the idea that less-than-lethal coercive force is perceived as having less "gravity" than lethal force. At the same time, the remote nature of many less-than-lethal technologies, a selling point for directed energy systems, does not equate this form of coercive force with close-in physical forms such as use of batons and truncheons.

To a degree, the "harmless" nature of less-than-lethal weapons technology in general has been oversold, resulting in a backlash when real-world results do not match the clean and bloodless claims. Although there has been criticism of CED employment, there is also the fact that these devices have been used quite frequently as an effective substitute for lethal force, saving lives in the process. Again, it should be stressed that less-than-lethal capabilities are a relatively new development for law enforcement. It is unfortunate, but doctrine and policy often spring from the aftermath of accidents and misuse. Not every real-world contingency can be foreseen. With all matters of coercive force, is it the tool itself or how it is used and who is using it that should be the primary subject of critical examination.

For some in the military the concept of less-than-lethal force is an alien concept, leading to difficult questions on the role of military force in operations other than war. As an instrument of foreign policy, the militaries of the United States and other Western nations, whether rightly or wrongly, are being employed in operations other than war. In the 21st century, civilian and military leaders have stressed the need to win "hearts and minds," if they were to achieve long-term victory. The United States has as of late found itself embroiled in multiple military operations other than war, where there is a need for civilian crowd control, as well as regular situations where distinguishing threat from innocents is difficult. Without less-than-lethal options, lethal firepower is often the only recourse, leading to locally tragic outcomes, as well as contributing to hostilities that would be counterproductive to the overall mission. However, in a war zone over self-restraint could result in one's own death and overall could also lead potentially to defeat. Traditional military weapons are meant to be lethal; meaning the avoidance of lethal force often translates into standing by while possibly allowing an enemy to attack. Military less-than-lethal weapons are envisioned as a third option: to employ coercive force to neutralize a potential threat but without necessarily killing the potential threat.

In the face of all the technical and political opposition to less-than-lethal DEWs, capabilities are slowly making their way into operational service. Though it was not used operationally, the U.S. Active Denial System (ADS) did make its way, briefly, into the war zone of Afghanistan in 2010.[46] ADS operates on the millimeter wave part of the EM spectrum, more associated with relatively safe radar and RF communications than the microwave-based cooking appliance found in most U.S. kitchens. Active millimeter wave technology is also used in some airport security scanners,[47] meaning U.S. frequent flyers may see more exposure to this part of the EM spectrum than anyone targeted by ADS in a warzone.

The broadcasts made by ADS are meant to penetrate less than half a millimeter into human skin, officially given as "1/64th of an inch"[48] to produce an "intolerable heating sensation."[49] This effect is meant to cause targeted individuals, or groups, to remove themselves from the beam. ADS avoids applying physical force or contact on the target, setting it apart from other less-than-lethal technologies, including water cannons.

ADS has faced controversy on many fronts. On one side there are those who have been arguing for immediate deployment since the moment ADS hardware was available. Often support for an accelerated ADS program is made alongside claims that it could have prevented some of the collateral loss of civilian life during U.S. operations at the time in Iraq and Afghanistan.[50] On the other side are those who oppose it due to its potential as an instrument of torture. An "intolerable heating sensation" combined with no avenue for escape could be construed as a form of torture in some opinions, including in cases where the lack of escape was purely accidental.

Proliferation could lead to this technology falling into the hands of regimes that would have less concern over causing pain and suffering, accidental or otherwise. The ability to cause discomfort, and possibly pain, without physiological effect is troubling as it is potentially torture without evidence. During the writing of this chapter, several popular uprisings in favor of democratic reform were occurring across North Africa and the Middle East, and less-advanced forms of crowd-control technology were employed by the regimes then in power with varying levels of success. During these confrontations there were many fatalities. Although it is uncertain if relatively benign forms of crowd control, such as ADS, are of interest to undemocratic regimes, they already have access to existing crowd-control technologies and do not seem quite as concerned over the injury potential of what they have on hand; the availability of such technology to them would be somewhat troubling to general U.S. and Western policy in support of democracy globally.

The particular application of millimeter wave energy employed by ADS is very new, meaning there is worry about its effects. ADS operates at 95 GHz,[51] which does technically mean it is operating in the broad range of the spectrum classified as microwaves,[52] which range from 1 to 300 GHz. Specifically, it fits into the International Telecommunication Union (ITU) classification for extremely high frequency (EHF), which is also a band commonly used for satellite communication, as well as WiMAX Internet access. NATO's system for RF classification would put ADS in the M-band, whereas the IEEE would call it a W-band system. EHF is also two bands higher than UHF radio and TV broadcasts of the past. Microwave ovens operate at less than 5 GHz, and in some cases at wavelengths technically below the microwave portion of the spectrum. Unfortunately these fine details have not prevented ADS from being confused with cooking technology.

Although all of the wavelengths involved are nonionizing forms of radiation, the photons involved do not have enough energy to knock electrons from atoms (thus making them very chemically reactive), there are persistent claims made about the damaging effects of RF communications (cell phones, wireless networking, radio) in the face of science consistently disproving such claims. On the EM spectrum, ionizing radiation, such as UV that is linked to cancers of the skin, is on the other side of the visible light from micro and radio waves. For that matter, heat (infrared radiation) is between microwaves and visible light.

Now it is true that ADS is not entirely harmless, it does officially have a "1/10 of 1% chance of injury."[53] Of the hundreds of test subjects who have been exposed to ADS, a handful have sustained minor burns, usually not requiring medical attention. So far the worse injury has been described as blistering.[54] One case of blistering requiring some medical treatment, and occurred due to accidental overexposure[55] out of what is still prototype software and hardware. In context, other less-than-lethal coercive technologies are also far from harmless, with consequences of exposure that include permanent maiming and death. Also, an effective less-than-lethal weapon does not include a requirement for complete lack of harm. However, despite the potential of this system, it has thus far been limited to being used on volunteers from the military, as well as members of the media, and those associated with the defense industry. It is of note that the military ADS program may not be the first operational use of this technology—contractor Raytheon is offering a scaled-down version to law enforcement and other domestic agencies for crowd control and infrastructure protection applications.[56]

Prior to ADS, there was great interest in using sound waves as a form of crowd control. Sound waves, as noted in the beginning of this chapter, are a form of energy. Technology such as long-range acoustic device (LRAD) by the LRAD Corporation is able to project sound in a very narrow cone to comparatively long distances, for a sound system. It is marketed as a communications device with an impressive ability to deliver a message to a specific target. LRAD technology, like any powerful speaker system, is capable of causing discomfort. Recently it has been employed by shipping to drive off pirates attacking shipping near Somalia.

Beyond simply projecting a very loud and uncomfortable noise, there has been long-term interest in using infrasonic sound, sound waves below that which humans can hear, to cause discomfort, disorientation and other coercive physiological effects.[57] So far much of this research has been inconclusive.[58] Another factor to consider is that the degree to which sound produces a physiological response differs for individuals. An effective weapon is one that produces reliable results, such as an "intolerable heating sensation." There are devices that produce a frequency of sound audible by a large percentage of teenagers, but inaudible to adults due to physiological changes with age. These systems are marketed to retailers in a bid to prevent teens from loitering around store fronts where it might be available, though these have come under some criticism due to their indiscriminate targeting of youth regardless of their behavior.[59]

As a less-than-lethal DEW, optical dazzlers have seen much more operational use. These types of systems include low-power lasers used both to get people's attention, in the context of warning them to slow down for a checkpoint, and to induce temporary vision impairment in tactical situations. For intentional tactical uses, laser dazzlers are useful in that they avoid accidental exposure due to their high degree of directionality—only someone on the axis of a laser beam will be affected by it.

Now wording is important on this subject, as there are treaty obligations restricting the capability of military dazzler lasers. The United States, as a party to the *1980 Convention on Certain Conventional Weapons* (CCW), is obligated to avoid weapons that are specifically intended to cause permanent visual impairment.[60] Specifically it is *Protocol IV, Protocol on Blinding Laser Weapons of the* CCW that limits the United States and other signatories from laser devices used in warfare, which are specifically used for permanent blinding. Essentially this means that between optical dazzlers and HELs there is wide range of energy levels and exposure mechanics that must be avoided as long as the treaty remains in force for the United States. Operationally this manifests itself in a combination of intentionally low power levels, electronic safeties (including systems to judge how much power is being delivered and systems for recognizing human eyes being in the line of sight), and training of operators.

This protocol does not restrict the use of weapons that may cause permanent blinding as a collateral or accidental effect, only systems where the desired effect is permanent blinding of human eyesight. Laser devices may still be used to damage optical-sensor equipment, opening potential loopholes for using optical-jamming equipment, which are not covered, as these devices are not specifically antipersonnel in nature. The proliferation of optical and heat-seeking missile threats has led to calls for civilian airliners to be equipped with countermeasures, including laser-based jamming systems that destroy a missiles ability to "see" the airliner. This means that potentially civilian airliners would be equipped with these devices, with potential for causing eye damage to those on the ground if they ever had to be used.

HEL weapons may, as a side effect, cause optical discomfort and injury from many of the atmospheric effects that inhibit propagation of laser energy to the target. Energy not delivered to the target has to go somewhere, and may be scattered in an unwanted manner. Air molecules may also absorb sufficient laser energy to briefly become incandescent. HEL weapons are not alone in their potential to cause collateral eye damage—some chemical reactions that may have potential as explosives generate damaging amounts of light, as do nuclear detonations. Again, these are collateral effects and therefore not covered by the CCW.

It is alleged that being a signatory to the CCW has not stopped other nations from procuring antipersonnel laser weapons capable of causing permanent blinding. Norinco, a Chinese company with a large portfolio of products, including many military goods, was for a brief time in the late 1990s marketing the *ZM-87 Portable Laser Disturber*,[61] a system some sources have cited as being sold with blinding as one of its antipersonnel capabilities, in addition to the less-controversial counter-optical sensor and temporary dazzling functions. In one article concerning laser illumination of U.S. helicopters by North Korean forces, the ZM-87, as one possible laser system possessed by the North Korea, is described as being capable of causing eye damage under some circumstances.[62]

Another aspect of using lasers to temporarily confuse sensors and human eyesight is the problem of accidental exposure. The proliferation of lasers has, however, led to a rapidly growing database on unintentional, nuisance, and malicious laser exposures. Every year there are hundreds of these incidents reported. This has included the illumination of commercial aircraft, with the result of impairment to the vision of the pilots.[63] It should also be stressed that these aircraft are at low altitude carrying out takeoffs and approaches for landing, meaning not only are these laser illuminations interfering with the safe operation of aircraft, they are doing so while aircraft have the greatest potential to hit the ground. Sometimes the laser illumination was found to be accidental, such as from light shows. At other times individuals have been prosecuted for illegal use of laser devices. Lasers able to project intense beams over long distances, such as relatively new battery-powered green light lasers, are freely available for under $100 in the United States, adding to concerns that the military may be far behind the general public in using lasers to interfere with vehicle operators.

Related to the use of lasers as a means to temporarily or permanently degrade human eyesight is the use of lasers as an ASAT weapon. The sensors aboard optical reconnaissance satellites are like the optical sensors on a missile, susceptible to temporary and permanent damage from exposure to relatively low-power laser illumination. During the Bill Clinton presidency a U.S. laser was used to illuminate a U.S. satellite as part of tests to assess satellite vulnerability, and sparked fears over U.S. intentions to deploy a laser ASAT capability.[64] Recently the Chinese have been accused of using a satellite-blinding capability against U.S. satellites, but like the previous example this may have just been another laser sensor test though one has to wonder why China chose to illuminate a foreign satellite if the test was intended to be of a benign nature.[65]

The CCW also does not prohibit the use of lasers with an intended lethal antipersonnel role. If the hundreds of kilowatts of laser energy delivered by ATL are capable of antimaterial effects, then surely it should be enough energy delivered to have some kind of lethal antipersonnel effect. In the antipersonnel role, a laser would have the potential advantages of range over other precision antipersonnel methods such as sniper rifles, which have to contend with gravity, air and wind resistance, and are limited in the end by ballistic technology. Being a new technology, there is also a potential for "novel" effects against a human target. At the same time, some of these novel effects could lead to criticism as opponents of such technology could make a claim that these effects would constitute a breach of international standards and norms. More than one article has indicated that there is a reluctance to discuss the potential lethal antipersonnel uses of DEW, despite there being no actual inhibitions (norms and treaty obligations) against it.[66] Like the counter-missile role, the speed of lethal effect would be important to determining whether a laser system would meet obligations on the avoidance of excessive injury.

Ironically, the use of DEW as simply a longer-range precision target engagement capability has the potential to save lives under the same rationale that defines the legitimate military role of snipers. In a way, DEW is the ultimate application of the precision-warfare paradigm, the ability to not just reliably bring specific pieces to equipment under attack from a standoff range, but the ability to target specific components, and troubling individuals. This last class of targets, specific individuals of concern, people who are enemy leadership and facilitators, such as terrorist bomb makers and financiers, takes DEW research into the controversial topic of targeted killing. This is a grey area in international behavior, and touches on the fluid delimitation between legitimate military activity and espionage. Some specific individuals are clearly military targets, such as uniformed commanders in times of war. However, others are not so clear; although a bomb maker killed in the act of constructing an IED would be the legitimate elimination of a clear and present threat, the terrorist financier is for many less of one.

A large-enough engagement range would allow a U.S. DEW platform to be hovering in international airspace, while the target may be residing well within an uncooperative nation. This is an operational concept that raises its own complications, including issues of sovereignty—uncooperative nations could include cases where time or other operation constraints did not allow for permission for the strike to be obtained. There is precedent for this type of action—as retaliation against terrorist outrages in the 1990s, the Clinton administration authorized cruise-missile strikes on targets with terrorist connections. More recently the United States has been using armed UAVs (see next chapter) to target terrorists in the Middle East and tribal regions of Pakistan that seem relatively uncontrolled by the Pakistani government, but has come under criticism that the small munitions employed still cause excessive collateral damage. A DEW-based strike weapon would offer an unparalleled level of precision, the ability to pick out only the guilty parties from a crowd. It must, however, be coupled with real-time targeting, though instead of a limitation this could be another advantage to this concept. Cruise missiles take time to strike—a laser or other form of EM energy utilized as the kill mechanism of a DEW would act almost instantly—meaning a targeting system with sufficient resolution to positively ID a human size target would also give immediate postattack analysis of its results.

It is presently clear that near-term deployment of higher-energy DEW capabilities is unlikely. Chemical HELs such as the ABL and THEL have conceptual difficulties, among them the need for a chemical reactants supply. Purely electrical HELs are a less-mature technology that is still years before reaching the power levels demonstrated by the troubled ABL and THEL programs. Problem-free development is practically unheard of in the creation any complex systems of systems. DEW development has faced additional challenges from both a checkered history, and perceived "newness." The lack of operation experience feeds both the overselling of capabilities and paranoia over its seemingly controversial aspects. These factors work against acceptance of DEW as a useful military technology outside of a few niche applications. However, it is these niche applications that may very well prove the robustness and other operational qualities needed in a weapon system.

Notes

1. IEEE Global History Network, "Theodore H. Maiman," http://www.ieeeghn. org/wiki/index.php/Theodore_H._Maiman.

2. Doug Beason, *The E-Bomb* (Cambridge, Massachusetts: Da Capo Press, 2005), 59.

3. "Charles H. Townes—Biography," Nobelprize.org, June 1, 2012, http://www. nobelprize.org/nobel_prizes/physics/laureates/1964/townes-bio.html.

4. Rüdiger Paschotta, "Beam Quality," in *Encyclopedia for Photonics and Laser Technology*, October 2008. http://www.rp-photonics.com/beam_quality.html.

5. Doug Beason, *The E-Bomb* (Cambridge, Massachusetts: Da Capo Press, 2005), 64.

6. Find press release, preferably USN for January 2011 breakthrough/contracts/ work on FEL.

7. Note that GPS guidance is cheaper, but less accurate due to inaccuracies inherent to GPS. For both the skill of the operator must also be taken into account.

8. Note that the first DARPA challenge resulted in no winners as no competitor was able to complete the course, the second did have several vehicles successfully navigate the desert, and the final one had vehicles successfully navigating an urban scenario.

9. Acknowledge possible incidents of intentional blinding by lasers, such as the Chinese ZM-87 portable laser disturber, and incident where Russian forces were suspected of using a blinding laser against NATO peacekeepers in Kosovo.

10. Missile Defense Agency, "Airborne Laser Test Bed Successful in Lethal Intercept Experiment," February 11, 2010, http://www.mda.mil/news/10news 0002.html.

11. Department of Defense, "DoD News Briefing with Secretary Gates from the Pentagon," April 6, 2009, http://www.defense.gov/transcripts/transcript. aspx?transcriptid=4396.

12. Ibid.

13. Federation of American Scientists, "Airborne Laser," http://www.fas.org/spp/ starwars/program/abl.htm.

14. William J. Broad, "U.S. and Israel Shelved Laser as a Defense," *New York Times*, July 30, 2006, http://www.nytimes.com/2006/07/30/world/middleeast/30laser. html?_r=1.

15. Northrop Grumman, "Northrop Grumman Laser 'Firsts'," http://www.as. northropgrumman.com/by_capability/directedenergy/laserfirsts/index.html.

16. Defense Science Board, *Report of Defense Science Board Task Force on Directed Energy Weapons*, December 2007, http://www.dtic.mil/cgi-bin/GetTRDoc? AD=ADA476320

17. Ibid.

18. Department of Defense. *Report of Defense Science Board Task Force on High Energy Laser Weapon Systems Application*, June 2001, http://www.acq.osd.mil/dsb/reports/rephel.pdf.

19. Defense Science Board, *Report of Defense Science Board Task Force on Directed Energy Weapons*, December 2007, http://www.dtic.mil/cgi-bin/GetTRDoc?AD=ADA476320.

20. Global Security, "Mobile Tactical High Energy Laser," http://www.globalsecurity.org/space/systems/mthel.htm.

21. Moore's Law specifically postulates that the number of transistors found in common (mass produced) integrated circuits (ICs) doubles every two years. Eventually physics will prevent transistor-based logic circuits from getting any smaller, though smaller transistors are not the only method to double numbers on an integrated circuit (bigger ICs, multilayered ICs, etc.).

22. Amos Harel and Avi Issacharoff, "Top Official: Israel Gave No Guarantees In Exchange for Gaza Truce," *Haaretz*, March 14, 2012, http://www.haaretz.com/print-edition/news/top-official-israel-gave-no-guarantees-in-exchange-for-gaza-truce-1.418328.

23. Northrop Grumman, "Northrop Grumman Chosen to Increase Efficiency for Next-Generation Military Laser Technology," September 28, 2010, http://www.irconnect.com/noc/press/pages/news_releases.html?d=202483.

24. United States Navy, "Office of Naval Research Achieves Milestone with Free Electron Laser Program," January 19, 2011, http://www.onr.navy.mil/Media-Center/Press-Releases/2011/Free-Electron-Laser-Milestone.aspx.

25. Ibid.

26. Ibid.

27. Boeing, "Boeing Advanced Tactical Laser Defeats Ground Target in Flight Test," September 1, 2009, http://boeing.mediaroom.com/index.php?s=43&item=817.

28. National Institute of Standards and Technology, "CODATA Value: Speed of Light in Vacuum," http://physics.nist.gov/cgi-bin/cuu/Value?c.

29. Nuclear warship power plants would of course remove this requirement.

30. Commission to Assess United States National Security Space Management and Organization. *Report of the Commission to Assess United States National Security Space Management and Organization*, January 11, 2001. http://www.space.gov/docs/fullreport.pdf.

31. U.S. EMP Commission. *Report of the Commission to Assess the Threat to the United States from Electromagnetic Pulse (EMP) Attack, Volume 1: Executive Report*, 2004, http://www.empcommission.org/docs/empc_exec_rpt.pdf.

32. Ibid.

33. Department of Defense, "Findings of the Nuclear Posture Review," January 9, 2002, http://www.defenselink.mil/news/Jan2002/g020109-D-6570C.html.

34. Wilson W. S. Wong and James Fergusson, *Military Space Power* (Santa Barbara: Praeger, 2010), 97.

35. David Fulghum, "For Now JSF Will Not Embrace Electronic Attack," *Aviation Week*, January 23, 2012, http://www.aviationweek.com/Article.aspx?id=/article-xml/AW_01_23_2012_p24-415796.xml.

36. Carlo Kopp, "The Electromagnetic Bomb—A Weapon of Electrical Mass Destruction," *Air & Space Power Journal*, 1996, http://www.airpower.maxwell.af.mil/airchronicles/cc/apjemp.html.

37. Ibid.

38. Department of the Air Force, Counter-Electronics High Power Microwave Advanced Missile Project (CHAMP) Joint Capability Technology Demonstration (JCTD), October 18, 2008, https://www.fbo.gov/index?s=opportunity&mode=form&id=e2daa9dccf59c9887810286dc9909d54&tab=core&_cview=1.

39. Ibid.

40. Boeing, "Boeing Awarded Contract to Develop Counter-Electronics HPM Aerial Demonstrator," May 15, 2009 http://boeing.mediaroom.com/index.php?s=43&item=656.

41. In November of 2002 two shoulder-fired missiles were fired at an airliner taking off from Mombasa, Kenya. Neither missile hit the airliner, which proceeded to Israel.

42. Doug Beason, *The E-Bomb* (Cambridge, Massachusetts: Da Capo Press, 2005), 57.

43. The usage of the word "kinetic" here refers to destructive, as opposed to "energy of motion," which is the physics definition that shows up elsewhere in this book.

44. Jim Garamone, American Forces Press Service, "CENTCOM Charts Operation Iraqi Freedom Progress," March 25, 2003, http://www.defense.gov/news/newsarticle.aspx?id=29230.

45. 'Less than Lethal'? The Use of Stun Weapons in US Law Enforcement, 2008, http://www.amnesty.org/en/library/info/AMR51/010/2008/en.

46. Noah Shachtman, "U.S. Testing Pain Ray in Afghanistan (Updated Again)," *Wired*, June 19, 2010, http://www.wired.com/dangerroom/2010/06/u-s-testing-pain-ray-in-afghanistan/.

47. Passive millimeter wave sensors actually work off of the energy radiated by all objects, including people.

48. Department of Defense Non-Lethal Weapons Program, "Active Denial System Fact Sheet," http://jnlwp.defense.gov/pressroom/adt.html.

49. Ibid.

50. Sharon Weinberger, "No Pain Ray Weapon for Iraq (Updated and Bumped)," *Wired*, August 30, 2007, http://www.wired.com/dangerroom/2007/08/no-pain-ray-for/.

51. Department of Defense Non-Lethal Weapons Program, "Active Denial System Fact Sheet," http://jnlwp.defense.gov/pressroom/adt.html.

52. David Hambling, "Microwave Weapon Will Rain Pain from the Sky," *New Scientist*, July 23, 2009, http://www.newscientist.com/article/mg20327185.600-microwave-weapon-will-rain-pain-from-the-sky.html.

53. Department of Defense Non-Lethal Weapons Program, "Active Denial System Fact Sheet," http://jnlwp.defense.gov/pressroom/adt.html.

54. John M. Kenny, et al., *A Narrative Summary and Independent Assessment of the Active Denial System, The Human Effects Advisory Panel*, February 2011 http://jnlwp.defense.gov/pdf/heap.pdf.

55. Ibid.

56. Raytheon, "Silent Guardian" online sales brochure, http://www.raytheon.com/capabilities/products/stellent/groups/public/documents/content/cms04_017939.pdf.

57. Alvin Toffler and Heidi Toffler, *War and Anti-War* (New York: Little, Brown and Company, 1993), 129–30.

58. Ian Sample, "Pentagon considers ear-blasting anti-hijack gun," *New Scientist,* November 14, 2001, http://www.newscientist.com/article/dn1564.

59. British Broadcasting Corporation, "Calls to Ban 'Anti-Teen' Device," February 12, 2008, http://news.bbc.co.uk/2/hi/uk_news/7240180.stm.

60. *Convention on Certain Conventional Weapons,* 1980, http://treaties.un.org/Pages/ViewDetails.aspx?src=TREATY&mtdsg_no=XXVI-2&chapter=26&lang=en.

61. Jane's Information Group, "Laser weapons (China), Defensive weapons," http://www.janes.com/articles/Janes-Strategic-Weapon-Systems/Laser-weapons-China.html.

62. Franklin Fisher, "U.S. Says Apache Copters Were Targeted by Laser Weapons Near Korean DMZ," *Stars and Stripes,* May 14, 2003, http://www.stripes.com/news/u-s-says-apache-copters-were-targeted-by-laser-weapons-near-korean-dmz-1.9753.

63. Federal Aviation Administration, "The Effects of Laser Illumination on Operational and Visual Performance of Pilots during Final Approach," June 2004, http://www.faa.gov/library/reports/medical/oamtechreports/2000s/media/0409.pdf.

64. Arms Control Association, "U.S. Test-Fires 'MIRACL' at Satellite Reigniting ASAT Weapons Debate," *Arms Control Today,* October 1997, http://www.armscontrol.org/act/1997_10/miracloct.

65. Warren Ferster and Colin Clark, "NRO Confirms Chinese Laser Test Illuminated U.S. Spacecraft," *Space News,* October 3, 2006, http://www.spacenews.com/archive/archive06/chinalaser_1002.html.

66. David Hambling, "US Boasts of Laser Weapon's 'Plausible Deniability'," *New Scientist,* August 12, 2008, http://www.newscientist.com/article/dn14520-us-boasts-of-laser-weapons-plausible-deniability.html.

CHAPTER 4

Computer Autonomy

Computer systems with greater autonomy, though not specifically robots, are perhaps the emerging military technology that has the most potential for near-term application. The increasing pace of technological development is now running into human limitations on speed, cost effectiveness, and overall capability. At the same time, hard limits for computer technology seem to be near. Although increased computer autonomy is in many respects welcomed, these systems are unlikely to escape the need for human supervision in the near term. This is especially true if the Western way of war continues to emphasize low collateral damage and flexibility.

In the long term, the potential for "thinking machines" and thinking weapons raise many questions. Some concern practical matters of doctrine—how these systems are to be used, and investment—how to achieve the types of computer autonomy desired by the military. Others are more philosophical and ethical, such as the question of whether reducing the human costs of war, both in financial terms and casualties, and whether the incentives to resort to military action increase for those that possess these technologies. The conjunction of practical and the philosophical concerns forms the legal discourse over autonomous military computer systems. There is already some concern over the legal status of warfighters making use of the latest in remote warfare.[1] Overall, the most critical questions being where and when this removal of the human element in war is appropriate, if at all.

The Problem with the Term Robot

Although there is common usage of the word "robot," there is quite a bit of dispute as to what the word should specifically be applied to. A remote control (RC) car is not often given the label "robot," but common police bomb-disposal robots are given just that label despite being just as reliant on human control to govern its actions. With munitions, "fire-and-forget" missiles and torpedoes are nothing new and under some definitions would qualify as "robots." Once unleashed, a "fire-and-forget" weapon will autonomously seek after a target based on its limited perception of the world. Despite its limited interactions with the world, and short autonomous lifespan, a smart weapon can be said to make more "decisions" about what it is doing than a remote-controlled machine such as an RC car or police bomb-disposal robot.

The word robot in the English language is attributed to the Karel Čapek play *R.U.R.*, where the word robot comes from the Czech word "robota." The subject of the play and source of its title, Rossum's Universal Robots, are biological constructions, more akin to the controversies over genetic engineering and other biotech of chapter six. The Czech word "robota" is variously translated to mean slave, serf, or one who performs drudgery. *R.U.R.* itself has an uprising by the robots midway through the play, perhaps being an allegory to slave/serf/peasant uprising. *R.U.R.* often comes up as a cautionary tale about artificial intelligence (AI) running amok, a media trope that has certainty affected public perceptions of real-world use of autonomous military systems.

Lack of a physical presence has not inhibited the use of the term robot. In many of today's more complex computer games, the player competes against computer-controlled agents, or "bots," that instead of following a preset pattern of actions, have some capacity to react to the player's actions, the game environment, and even other "bots." An antivirus program struggling on its own between updates to detect new and unfamiliar computer viruses also falls under a generalized concept of "robot." These limited forms of autonomy exist only within a computer's memory. These two examples are autonomous agents that have a capacity to perceive, make decisions on its actions, and interact with its environment—these agents and the environment all being purely information in the form of computer code and data files. At its core, the ability to make decisions is what sets a true robot apart from other forms of automation.

Now any random-number generator can produce decisions of a sort, so it is important for the autonomous computer system to be able to make, if not the perfect decision consistently, reasonable decisions based on the supplied data. The speed, accuracy, and capacity of computers to process data, as long as it is in a form it can use, has led to great interest in such decision-making systems. Getting ahead of the competition in the decision-making cycle is applicable to both warfare and business. Among the more popular decision-making models, USAF Colonel John Boyd's OODA (observe, orient, decide, act) loop places a premium on getting ahead of an opponent's ability to make decisions. This is true in the world of business, where expert systems and other forms of limited AI are making stock-market transactions, reacting to events faster than human traders could.

The limited autonomy found in some antivirus software tries to uncover computer virus through recognition of virus behaviors and characteristics. In theory this allows the antivirus to detect viruses newer than its last update. These programs, if the user allows, have the capability to decide on their own what files could be threats and take action on their own. Computer viruses represent an arms race where simply trying to keep up with new virus and variants of existing virus would always leave the antivirus vendor playing catch-up. This capability is, however, also capable of misidentifying, and then deleting, perfectly harmless files, including those critical to a computer's normal operation. This form of autonomy is often an optional capability used at one's own risk as mistakes can and do happen with antivirus software deleting innocent critical files. This is the computer equivalent of collateral damage.

At the far end of autonomous systems research is the nebulous concept of "strong AI." Strong AI is often described as the point where machine intelligence meets or exceeds human intelligence and ability to reason. The definition of strong AI, however, often omits the problematic definitions of what constitutes human intelligence and ability to reason. It is a growing area of philosophical, theological, and legal debate, as the concept is basically the creation of artificial life; if not, then a form of artificial sentience. The United Kingdom has in a recent series of scientific reports on emerging technologies studied the concept of AI rights.[2] Strong AI when portrayed by mass media usually warns of it deciding to destroy humanity.

In the short and medium terms, the challenges and promise of strong AI are unlikely to be factors in real-world military autonomous systems. Although it may be useful at some time in the future to create an artificial soldier with all the versatility of the human mind and form, it may not be necessary to take it to the level of strong AI. The point of many autonomous systems is to produce inhuman endurance and precision, not to produce machines capable of reciprocating feelings of loyalty and fellowship, as some soldiers have expressed toward their unit bomb-disposal robot.[3] At present it remains a challenge for AI to recognize humans in general, let alone abstract concepts such as "loyalty" or "harm," leaving seemingly compelling problems such as Asimov's three (sometimes four) laws of robotics[4] conflicting with the notion of military robots as fanciful debates for years to come.

Rapid advances in basic computer technology are leading toward computer systems that rival some measures of the human brain (keeping in mind that the function of the human brain is far from completely understood).

The convergence of many of the technology areas covered in this book, es-pecially when combined with traditionally high levels of military and intel-ligence-agency funding, leave many strong AI proponents hopeful. Some, such as prolific technology writer Douglas Mulhal, even speculate that strong AI may emerge unintentionally out of future military or intelligence-agency projects.[5]

What Is a Computer and Why Everyone Has Great Expectations

The word computer means different things to different people. Conver-gence with telecommunications has given midlevel cell phones more pro-cessing power, and arguably versatility, than desktop computers that were commonly available at the turn of the century. Essentially what is com-monly thought of as a computer is a general-purpose electronic device, able to load a variety of programs to perform functions involving the retrieval and manipulation of information. In contrast a desktop calculator and very basic cell phones are limited-function electronic devices; despite sharing many of the basic operating principles such as digital logic and integrated circuit technology, they cannot be easily repurposed for more than what they were manufactured for.

Modern computers are based on digital logic—mathematical operations based on finite values.[6] Specifically everything done by a modern computer is eventually reduced to the storage and manipulation of numbers in a two-digit (binary) numbering system, the 0s and 1s now ingrained into popular culture's concept of computers. Boolean algebra, a form of mathematics specialized for a two-value system, 0s and 1s, on and off, true and false, is the underlying principle of digital computing. In terms of hardware, any Boolean function can be produced by a combination of NAND (negated AND) logic gates, meaning that a sufficient number of NAND gates produces a system capable of handling all mathematical functions. The details of the architecture used in real-world computer chips are more complicated than simply having very large collections of NAND gates, with specialized logic circuits to optimize performance; however, in theory, given no requirements for speed, size, and power efficiency, all the calculations performed by modern computing could be eventually performed by a very large array of NAND gates.[7]

Logic gates are built up from transistors. A transistor in this usage acts as a two-state switch used to represent the binary digits of 0 and 1. The perfor-mance metrics of a modern computer element, such as a CPU or memory chip is related to the amount of transistors found in the component, and the speed at which the transistors can operate. Transistor technology has evolved from a collection of different metals soldered together to thin lay-ers of metal and rare earth elements layered onto silicon wafers. Candidates for future transistors include complex arrangements of carbon atoms (carbon nanotubes and graphene) and self-assembling DNA molecules.

Prior to the modern digital computer, there were machines called analog computers. Analog systems are based on physical quantities along a continuous scale, where there is a smooth transition from point to point—an infinite number of different quantities between points. Digital representations only have a finite number of possible values. Now it is argued that with only the nature of the physical medium limiting resolution, there is a richness found in analog representations, old fashion photography and analog audio, not possible with digital formats. Digital technology is limited to the resolution of the original impression captured. With traditional photography an image can be easily enlarged to very large sizes. Digital photography if enlarged beyond the limits of the original image suffers from pixelation (becoming a series of jagged square blocks).

The advantage of digital encoded data is that such data can be easily stored, manipulated, and transmitted. Imagery intelligence gathered by satellite or aircraft can be immediately accessed as many times as needed. Advances in data storage have meant that very high-resolution data can be stored in a physically compact form. Extra information is easily attached or embedded without damage to the original data. In the past film would have be recovered and processed, meaning it took time to reach the analyst phase and rarely was supplied to lower echelons of command. Recovery of film is not an easy task—aircrafts and drones can be shot down, and satellite film-return systems involved the challenges of de-orbiting and reentry from space. A physical medium implies physical limits on how much can be recorded on a mission. Copying an analog medium introduces errors. Photographic intelligence can suffer from being overly marked on, or otherwise damaged, during analysis.

The contemporary technological era, the Information Age, unlike the prior Atomic and Space Ages, has its foundation embedded in everyday life. For the military these technologies have manifested in net-centric warfare, the current Revolution in Military Affairs (RMA), and other superlatives to describe the modes of Information Age warfare spearheaded by the United States. Part of the reason that computer technology has become a dominant factor in contemporary society has been the exponential growth in computer power. Moore's law, named after Intel cofounder Dr. Gordon Moore, posits that, "The number of transistors incorporated in a chip will approximately double every 24 months."[8] The timeframe of 24 months was actually an update made in 1975; earlier in 1965 Dr. Moore expressed that doubling would occur every year.[9] Others have placed the recent the rate of doubling at every 18 months.[10] The greatest factor in this power increase is due to continuing advances in integrated circuit production; "Cramming More

Components onto Integrated Circuits" was the title of Dr. Moore's 1975 article updating Moore's law.

The continuing exponential increase in computing power has been a strong factor in the development of many of the emerging (military) technologies covered in this book. Compact computing power will be needed for many of the on-orbit applications that justify the development of ubiquitous military space access. Computer control is critical to many of the technologies needed for a viable long-range directed energy weapon (DEW) attack; examples of such technologies include atmospheric compensation and the basic sensor-to-shoot cycle for engaging targets at thousands of kilometers of range. For nanotechnology and computer power, it is clearly a two-way relationship; nanotechnology is one possible technology to continue the exponential growth of computer power, and the computer power being made available via Moore's law is needed in modeling the properties of atomic-scale constructions. The exponential growth in computer power is credited as being the key factor to completing the map of the human genome,[11] and will continue to be an important tool in finding applications for genetic research and other biotechnology breakthroughs. Advancements in understanding how the brain functions, and learning in particular, are serving as models for solving many of the challenges that keep general-purpose machine autonomy out of reach.

Now that there will be limits to how far a transistor may be shrunk, nearing the atomic scale, quantum physics introduces many complications beyond the scope of this book. Then again Moore's law does not impose a size limit on the conceptual mass-produced integrated circuit; if transistors cannot get smaller, then numbers could simply expand outward (larger microchips), and upward (moving beyond two-dimensional architecture to multilevel three-dimensional architectures).

This wealth in computer power coupled with rise in the Internet has brought about many changes in society, with many surprising results. At the same time there was much hype over the Information Age, much of it predictable in hindsight. The fortunes lost during the dot.com bubble burst are warnings that simply throwing money at trendy-sounding technology is no guarantee of wealth. Post burst, Web 2.0 has, however, been more successful in leveraging technology as a new medium for human interaction. Information revolution success stories include Internet commerce for real products, as well as turning online interaction, such as online search,[12] into a commodity. These later Internet successes were not even imagined in 1990s when public Internet access was still emerging, and largely regarded as a toy for the small technically inclined segment of society. Today some jurisdictions are treating Internet access as more of a necessary utility to modern life, and in some cases as a right.[13]

In the security and defense realm, Information Age warfare has also seen hype, some justified, others not so much. The 2003 U.S.-Iraq War can be, and is, argued as being an example of both. RMA and net-centric warfighting concepts were used to rapidly overcome the Iraqi military and initially oust Saddam Hussein. The long struggle to establish security in the aftermath has for many raised some doubts over the hype over RMA and net-centric concepts (notwithstanding the fact that these concepts were originally meant for a different type of conflict than the low-level conflict seen in Iraq after 2003). A major theme in P. W. Singer's *Wired for War* is the contention that the early hype and interest over Information Age warfare was misdirected on computer networks, and the application of corporate information technology strategies (some of a questionable nature as many businesses that exemplified such strategies suffered greatly during the dot.com bust) to the military,[14] instead of on the robotic warfare systems that Information Age technology was slowly bringing into reality. Another theme was the proliferation and, perhaps more importantly, increasing acceptance of battlefield robotics, specifically land-warfare systems; this effectively was an argument that the true RMA of the Information Age is the removal of people from the platforms allowed by networks and perhaps later by autonomous computer systems.

> It is far more important that humans' 5,000-year-old monopoly over the fighting of war is over.[15]

The networking of computers and computer-controlled devices has led to other areas of security and defense thinking on the Information Age: cyber security and cyber warfare. Computer virus writing started off as pranks and experimentation, but today are means for serious criminal activity. The U.S. government's Internet Crime Complaint Center, a partnership primarily between the Federal Bureau of Investigation (FBI) and National White Collar Crime Center (NW3C), noted in its 2010 report a growing diversity of Internet crimes being reported.[16] Many of the same tools used to hijack, steal information, and otherwise attack Internet resources can also be directed toward political ends.

Incidents of interstate cyber conflict have been clouded by the ease at which an attack's point of origin can be hidden. By hijacking the services of multiple computers across multiple networks attackers can mask their trail.[17] Often it becomes difficult to discern if the malicious activity is the result of criminal entities on their own, criminal entities employed by foreign governments, or computer-warfare organizations, such as those known to exist in the People's Republic of China.[18] Much of this activity is espionage, with U.S. government agencies and defense contractors as natural targets. The globalized nature of the Internet and the transnational nature of many computer components have raised not only the specter of foreign attacks breaching the thus far limited defenses in use, but also foreign intelligence and military organizations being able to preinstall backdoors into the software and hardware of critical U.S. military, intelligence, and civilian systems.[19]

In 2010, the Stuxnet computer virus made headlines by possibly being specifically targeted against industrial equipment connected to Iran's nuclear fuel enrichment program. The complexity of the Stuxnet computer virus has some analysts speculating that it could only have originated from the resources of one or more national governments who have been attempting to stop Iran's nuclear program, which is feared as a cover for nuclear weapons development.[20] In other words, Stuxnet potentially represents a case where a computer attack was an instrument of "politics by other means." If this assertion is true then for some analysts a Pandora's box has been opened, where all nations with computer-controlled infrastructure may come under threat. On the other hand the existence of vulnerable computer-controlled infrastructure, largely a result of human inaction, is also an invitation in itself for attack. At the time of writing the origins of the Stuxnet computer virus is unknown, but some of the specific techniques it employed as a virus are now being found in other forms of malicious software, this time of a criminal nature.[21]

Policies to counter cyber threats have been slow to emerge, in part because the potential of the Internet is not well understood. An increasing amount of business resources and industrial supervisory control and data acquisition (SCADA), are becoming accessible online, usually to increase productivity by allowing staff to work from any location with Internet access. A specific point of concern is the use of SCADA systems in the operation of critical infrastructure such as power plants. Underscoring this has been the fact that many of the recent incidents of cyber intrusion are less about sophisticated computer code and instead more about psychology and preying on people's lack of understanding and will to follow sensible computer-security processes. Security technologies such as encryption and Virtual Private Network (VPN) software, as well as hardware-based security devices—all rendered useless by improper use by employees. Inattention to locking down online resources opens vulnerabilities for attackers to stumble across. Social engineering uses subterfuge to convince end users to betray usernames, passwords, and other information useful for someone wanting to break into a computer network.

The power of computers has also allowed those who choose to betray the trust placed in them the capacity to cause greater damage, such as the case of U.S. Army Private First Class Bradley E. Manning who at the time of writing is alleged to have been the party responsible for copying and later delivering to unauthorized persons thousands of classified documents and videos. The sheer number of files involved introduces an indiscriminate element to the alleged actions; the stolen material covered diverse topics and activities of the U.S. government, with often only the restricted classification, and being shared on the same secure network being major commonalities.

There is of course an element of hype to war in cyberspace. Thus far there have been incidents of fraud, theft, and espionage. There is, however, nothing yet approaching the "digital Pearl Harbor" feared by many. Prior to the release of the July 2011 U.S. Department of Defense Strategy for Operating in Cyberspace, there was speculation that the new strategy would formalize the option for a "kinetic" response to a major cyber attack.[22] State-versus-state cyber attack, like an antisatellite attack, must be considered in the broader sense of what it is intended to achieve. It would be unlikely that such an attack would be pursued purely for its own effects, and more likely would be in support of some other military action. Moreover the attention and hype over the threat of cyber attack has been leading to constructive discussion and, importantly, action on methods to blunt attacks.

Another emerging aspect of Information Age international relations and international security is the use of social-media technologies in motivating political action. At the time this chapter was being written, popular uprisings demanding democracy and accountability in government were occurring across the Middle East and North Africa. These uprisings, referred to as both the *Arab Spring* and the *Jasmine Revolution*, were partially coordinated with social-media tools such as text messaging and online chat. Repressive regimes have been mindful of these developments, taking action to block access to these services and online resources.[23] Similar Internet-fuelled discontent in 2010 over charges of election fraud in Iran was suppressed by Iran's repressive regime. Social media, including the easy ability to post digital videos online, are allowing the world a view into such events, when traditional journalists are denied access. The events of 2010 and 2011 will certainly provide many lessons for both those who seek freedom and the regimes that seek to maintain a grasp on power.

A darker side to technology's capacity to inspire action is that it is also being used by violent fringe groups to recruit new members and distribute training material and propaganda. Although not able to mobilize the numbers seen in the protests demanding democratic reform in the Middle East, the nature of terrorism only requires a handful of fanatics to have an effect. This has not gone unnoticed by Western intelligence and law-enforcement agencies. Reportedly a web magazine run by Islamic militants was recently hijacked by the United Kingdom's Government Communications Headquarters (GCHQ), replacing bomb-making instructions with cupcake recipes.[24]

Digital Realities

It is reasonable to expect computers to continue to grow in their capacity to handle numerical challenges for quite some time, but warfare is never a clean numbers problem. Over time computers have been able to carry out tasks in increasingly complex environments. Though mastering the game of chess may be daunting, it is relatively simple in comparison to navigating a personal home. The average household has an untold capacity for clutter and change is many magnitudes more complex than a chess board with only 64 squares, 32 pieces, and small number of rules for movement. Computers and people, for the most part, find the opposite of these two activities challenging—not everyone can play chess, but nearly everyone can find their way around a home. Computers have been able to offer a challenging game of chess before the advent of personal computing, but developing a commercially viable computer system for navigating a home remains a difficult challenge for robotics.

Presently the chaos of the real world is largely avoided, still allowing for some forms of autonomous operation. Most models of robotic vacuums available today dispense with perceiving and memorizing the entire room, let alone developing a plan of action specific to the room it is in, and instead act on a combination of random pattern running and the ability to detect and avoid obstacles and drop-offs in its immediate vicinity. By not having a memory, or needing to plan, and simply reacting to immediate sensor inputs as it attempts to run a series of patterns designed to maximize floor coverage, the robot vacuum deals with the real world by being ignorant to it outside of immediate concerns, such as avoiding drops and slowing down when nearing a wall. Cutting out the need for cognition, or anything approaching strong AI decision-making capabilities, was for a time controversial in robotics circles. Dr. Rodney Brooks prior to iRobot success noted that his early work on robotics without cognition amounted to a form of heresy.[25] Sidestepping the problem of robot perception is also to a large extent how modern autonomous unmanned aerial vehicles (UAVs) or unmanned air systems (UASs), as these machines are now termed, can navigate global distances. Within the volume of space where the NavStar global positioning system (GPS) is usable, the coordinates for any point in space can be determined within a few meters. The reality of easily obtainable, constant, and accurate knowledge about one's own position has greatly simplified the problem of autonomous navigation. GPS-guided machines in effect do not operate with knowledge of real-world geography.

Prior to GPS, the options available to autonomous vehicle operations included terrain contour matching (TERCOM) and later digital scene-mapping area correlator (DSMAC). Both TERCOM and DSMAC are dependent on geospatial data of the ground under the flight path and machine perception able to recognize these "landmarks." Systems that make use of star tracking are also in use, notably in missile and spacecraft applications. These earlier methods for autonomous navigation attempted to replicate in a machine some of the capabilities of a skilled navigator to recognize landmarks. Inertial navigation systems (INS) are built around precision sensors tracking a vehicle's own movements to determine where it is relative to its origin.

The limits of computer technology at the time, however, meant that major deviations from the planned flight path, or inaccuracies in the preloaded maps, would likely result in the missile or drone losing its way. The technology of the time also required extensive preparation of geospatial data needed for a mission, and could not be changed in-flight, or at the last moment. INS on the other hand losses accuracy with distance and time as minute inaccuracies in even highly refined gyroscopes and accelerometers build up. Without the situational awareness of a skilled crew onboard to direct a vehicle closer, the applications for these forms of autonomous flight were limited. Early TERCOM and DSMAC were common navigation systems found on long-range nuclear-armed cruise missiles, where accuracy of a few hundred meters was well within what was needed for these to be effective weapons. Despite these limitations, in some ways, these earlier weapons had more "robotic" characteristics in that they had to be somewhat capable of sensing real-world surroundings.

Simply being supplied with a vehicle's present coordinates reduces the navigation problem to a matter of getting from one set of precisely known coordinates to another. This is one factor in the proliferation of all-weather precision weapons, such as the Joint Direct Attack Munition (JDAM) series. Though limited (often absent) situational awareness does present problems, especially with regard to operating in crowded airspace, GPS navigation has also made possible long-range autonomous flight. On August 24, 2001, a U.S. *Global Hawk* UAS named *Southern Cross II*[26] landed in Australia, successfully completing a milestone autonomous flight that originated from Edwards air force base in the continental United States.[27] Less than two years later, on August 12, 2003, *TAM-5*, otherwise known as *The Spirit of Butts Farm*,[28] became the first model airplane[29] to cross the Atlantic Ocean nonstop, a feat performed with GPS navigation on autopilot for the majority of its flight.[30] The latter record is an indication of the ubiquity of GPS, and the very low cost associated with this form of autonomy.

Converting the real world into data that a computer can comprehend is among the greatest challenges facing autonomous systems today. On a mobile autonomous system this of course has to be done quickly enough for the machine to act upon what it is able to perceive. As technology develops, the tracking of subtler signatures becomes possible. Early forms of machine perception, from well before the digital-computer revolution, could perceive and track high-contrast events such as a jet engine's nozzle, although as the air war over Vietnam demonstrated, the sun also provided a seemingly viable target. Today, there is work inside the laboratory, and among at-home computer enthusiasts, to refine computer software able to detect eyes and faces in pictures and video inputs. Commercial uses for this later technology include using eye detection to cue and focus cameras, reducing "red eye" in photography, and biometric security (not simply detecting an eye, but also unique identifiers of a specific eye).[31] Autonomous face recognition and identification software used by the Internet company Facebook[32] and others[33] have raised privacy concerns, highlighting that the ease of using new technology has outpaced the governance of such capabilities. Biometric technology is being combined with cognitive and behavioral research in hopes that it can be an early-warning tool for terrorist and criminal acts.[34]

Face recognition, for identification purposes, expands on simply locating face-like images and collects biometric data, such as proportions and distances between features, and compares these numerical data points against a database. The latest smart weapons similarly are able to identify targets from a small onboard database of signatures corresponding to possible targets.[35] This is something of a brute-force method, as expanding the repertoire of things to be identified means a larger library of known sensor inputs, coupled with faster processing and even additional specialized recognition algorithms to use to conduct the search of this database. This rapidly becomes an insurmountable data-entry challenge.

General object recognition is still something being studied, both in how humans and animals are able to perceive objects, and in how to either replicate or approximate such perceptive capabilities in machines.[36] Although very basic computer logic can track an aircraft based on tracking specific sensor inputs, having a computer recognize the concept of an "aircraft" from general-purpose sensory data is still early in its infancy—an apt description in that that many of these efforts are modeled on the growth and development of a child's ability to perceive of the world.[37]

The chaos posed to machine perception in the relatively benign setting of everyday United States (or the Western world in general) is far removed from the chaos that is warfare. Apart from the general perception problem, the target may be actively trying to evade the autonomous weapon. The early heat-seeking missile could only discern infrared energy differences, heat, detected by its sensor, and has no way of knowing if the heat source is a jet engine tailpipe, a flare, or the sun. Current-generation heat-seeking missiles employ a range of processing and target-discrimination techniques that have not only been able to discern between the target, decoy, and the sun, but have also, in recent times, been able to recognize the target without actually seeing it at launch, otherwise called lock-on-after-launch (LOAL).

Now mistakes will happen sooner or later, but the computer is sometimes regarded as having an advantage in that there is an expectation that a computer will reliably replicate the same correct decisions given the same inputs. However, this reliability with following orders becomes a liability in light of the limitations of sensor technology. Given similar inputs, it may consistently produce the same wrong decisions. A valid target, a friendly military platform, and a civilian vehicle could all easily present an identical signature as far as a particular sensor may be concerned. Combining multiple sensor types is one technique to narrow down the selection. Military and civilian vehicles may have identical signatures to some sensors; for other sensors, the signatures could be quite distinct.[38] Identification friend or foe (IFF), systems that broadcast on-demand coded ID, is also useful, though with respect to discussions on autonomous systems, examples of IFF failures and limitations do seem to come up often.[39] Then again, in the chaos of war humans are also liable for "friendly fire," and the use of autonomous systems to govern the release of weapons is such a recent (and controversial) development that it remains to be seen if it is humans or computers that are better at preventing these incidents.

Decision Making

Connected to these problems of machine perception are the problems of machine understanding of syntax and context. At least fiction has gotten the tendency for machine intelligence to be portrayed as overly literal somewhat correct. Giving machines a better understanding of context and syntax would make computers increasingly capable of handling real-world situations where the appropriate response to stimuli depends on an awareness of the situation it is in. Like pattern recognition in advance machine perception, solutions to the problem again include brute-force techniques such as attempting to program in contingencies for likely scenarios, to trainable systems that can be taught what an appropriate response is, and combinations of the two.

Most present day, and all historical, computer control systems can only simply respond to stimuli, and consequently there is a sentiment that this blind obedience would lead to tragic mistakes when combined with weapons. A human being on the other hand should be capable of more than blindly following orders, and is consequently argued as producing better quality of results, even under the stress of combat. One example of this would be Stanislav Petrov, who as a Soviet air defense lieutenant colonel on September 26, 1983, correctly dismissed reports from a satellite-based early-warning system that the United States had first launched one, then later a small salvo of intercontinental ballistic missiles (ICBMs) against the Soviet Union.[40] In some circles at least,[41] these decisions are credited with preventing an accidental nuclear war during a particularly tense period of Soviet-U.S. relations. This could also be an example of the "man-in-the-loop" needed to supervise computer-controlled systems. The Russian rebuttal to the acclaim being bestowed on Stanislav Petrov in 2006 contends both U.S. and Soviet/Russian nuclear command and control had, and continue to have, several people-in-the-loop just for this very purpose.

In the previous example, the early-warning system did what it was designed to do, warn of nuclear attack if its satellite-based sensors were set off.[42] It had no ability to consider that a single launch didn't make sense in terms of global nuclear war, or for that matter that if it made one mistake with its first warning that its second warning could equally be wrong. In this sense the duty officers, such as Stanislav Petrov, were there to fill in the context of the situation, including reasoning that the small number of reported ICBM launches was more indicative of an error, than of thermonuclear war. It and other computer-based early-warning systems of the time, and those in use today, are basically interfaces for early-warning sensors. Now of course interfaces filter data, which raises a problem familiar to management and leadership: that of the influence of information gatekeepers.

The still limited ability for a computer to determine context has many repercussions to how autonomous systems can and should be used. As mentioned earlier, it is possible to apply face-recognition software to large databases of digital photographs to help identify people for social purposes. However, basic face-recognition algorithms cannot tell if the face identified is a subject of the picture or someone in the background, and certainly cannot tell just from a digital picture if the person identified wants to be identified, or even wanted their picture taken in the first place. There are serious privacy implications from the idea that companies would simply presume anyone who is digitally photographed wants to be identified by their software unless they opt out.

For military autonomous systems, this would be moving from simply being able to recognize a target, to being able to determine when it would be appropriate to attack it. As one rational for autonomous weapons is the potential to reduced collateral damage, understanding the context of a situation where the target is found would likely involve some form autonomous evaluation of the rules of engagement (ROE) before carrying out an attack. ROE has become something of a controversial topic lately with rogue state and terrorist opponents openly attempting to evade attack by collocating their assets among noncombatants in a bid to both avoid detection and, failing that, using the very presence of civilian bystanders as shields, knowing that U.S. and Western forces have inhibitions against imposing excess collateral damage. Sometimes the ROE means allowing an enemy to escape to fight another day. Other times the value of the target means that in accordance with the ROE, the attack must be conducted, resulting in civilian collateral damage. Neither decision is easy to make, especially under the stress of combat. In the aftermath, these types of decisions are open to second guessing.

This concept of machines being able to recognize not just what to attack, but when to attack is controversial. The point made by the Russian press release is that nuclear weapons during the 1983 incident and to this day are too important to be placed under any kind of automated control (notwithstanding rumors of a parallel Soviet-era doomsday system[43]). Autonomous conventional warfare is, however, a different matter. Right now the decision of when to attack a target is done manually, a weapon is launched and all the shooter can do is wait for impact—being present on the launching platform or remotely controlling it does not change this. The responsibility for firing and potentially any consequences from firing rest with a shooter. In the case of the remotely controlled/supervised platform this is the man-in-the-loop that the Pentagon repeatedly states will be in charge of weapon release by unmanned platforms.

Now related to the controversy over whether AI is to be trusted with giving the decision to fire, is the matter of who right now is making those decisions. The Central Intelligence Agency, despite its usually secretive nature, has become quite prominent in its armed drone operations against terrorists in the Middle East, Afghanistan, and Pakistan. The major concerns cited in a recent United Nations Human Rights Council (UNHRC) report that included concern on drone warfare was the lack of transparency inherent to intelligence organizations.[44] While acknowledging that the drone attacks may be legitimate in themselves, the report brought into question the legal and organizational framework under which they are conducted.[45] The subject of collateral damage from prolific counterterrorism operations just adds to the legal discourse over matters of accountability and responsibility for firing.[46]

These questions of accountability and responsibility are being applied to the still-hypothetical autonomous system capable of making the decision to attack. As target identification is more developed than the capability to evaluate the situation, there are also fears that these weapons will be employed without a rigorous ability to consider collateral damage. In this case collateral damage would not be by mistake, but simply due to an undeveloped or oversold feature. Even without a restrictive spending climate, weapons' vendors have been accused of overselling features—this may someday include a weapon's ability to distinguish valid targets.

The basic antipersonnel landmine is a weapon abhorred by much of the world as an indiscriminate killer. It also happens to be an autonomous weapon, in that it is left behind with the expectation that it will go off once certain sensing criteria are met. Among the major objections to landmines is this type of weapon's general nonconformance to the, "principle that a distinction must be made between civilians and combatants."[47] The pressure switch of an antipersonnel landmine does not know if the foot setting it off belongs to a soldier or a civilian.

Yet at the same time landmines are a cost-effective means to deny territory or shape the movements of the enemy. Unlike many of its allies, the second argument has prevailed in the United States in its decision not to sign the *Convention on the Prohibition of the Use, Stockpiling, Production and Transfer of Anti-Personnel Mines and On Their Destruction* (also called the *Ottawa Treaty*). The United States is not alone in this regard, as many military powers such as Russia, People's Republic of China, and India have also declined to join the landmine treaty. Therefore it must be asked if, and when, limited discriminatory capability diminishes an autonomous weapon's utility as a precision weapon? Certainly, some ability for target discrimination would be better than the total lack in basic landmine, but how much is needed to address legitimate concerns.

It must also be remembered that someone in authority should know, and is responsible for, where the autonomous system is deployed. A tactical commander could include a weapon's cognitive limitations in the selection process for where and when to use a particular autonomous system. In this sense the use of an autonomous system again could be regarded as nothing more than firing a weapon, with the weight that it carries, except the weapon has the option of not attacking a geographical location if the target is not present or the situation has changed since the decision to fire. The true question then becomes whether the autonomous system's ability to distinguish targets is appropriate for the situation.

A related, somewhat philosophical problem is whether having weapons with some capacity for self-restraint over when and how to attack is an invitation for the use of these weapons. Does the weapons capacity for precision and limited harm, give a false sense of comfort, reducing the perceived moral and ethical weight involved with weapons' employment? Arguments are being made that, for similar reasons, less-than-lethal weapons produce a relaxed attitude toward their employment, leading to overuse.[48] Similarly there are charges that existing armed UAV capabilities, the ability to destroy targets without exposing one's own personnel to the immediate risks of attacking, are proving too tempting not to use.[49] It could be argued that if it were not for contemporary risk aversion toward casualties and collateral damage, these targets would have been attacked via pre-RMA methods, such as area bombing. Capabilities such as unmanned platforms, weapons of increasing precision, and platforms able to autonomously take advantage of brief opportunities to attack fill in for the firepower lost due to current attitudes on reasonable use of military force.

Despite efforts to limit collateral damage, there will always be critics to the use of coercive force. Although their humanitarian sentiment is to be commended, it sometimes seems that for some there is no amount of adequate risk and collateral-damage prevention short of complete abstinence from military force. Indeed there are limits to how much more collateral-damage prevention is sensible. The rise of the smart bomb in the 1970s and 1980s was to increase conventional weapons effectiveness. Part of this was to develop conventional-warfare options to avoid the many conceptual, operational, and even moral problems of tactical nuclear war in the face of the Soviet Union's numerical superiority. After success in the 1991 Gulf War, there continued to be funding for new generations of even smaller yet more accurate weapons, now with collateral-damage reduction as a stated goal. The smaller the target, the more expensive the capability needed to hit it. The sad reality is that human conflict is not going away, and increasing the level of precision has diminishing returns. These two factors mean death and maiming from war are not likely to be going away. However, the United States, the West, and even near peers are at least investing in capabilities that give the choice of a more humane form of warfare.

Operation Neptune Spear, the mission on May 1, 2011, to kill or capture Osama Bin Laden, the terrorist mastermind behind the 9/11 attack, highlights several limitations of contemporary unmanned warfare options. After an intense hunt lasting almost a decade (and earlier attempts to find him for other terrorist atrocities prior to 2001), Bin Laden was located hiding in "plain sight" in a relatively affluent neighborhood of Abbottabad, Pakistan. Reportedly the initial option considered was to simply destroy the compound that Bin Laden was hiding in with a precision air strike.[50] An air strike, however, not only risked significant collateral damage, but also had the problem of confirming that Bin Laden was in the compound in the first place. The terrorist leader had been careful to avoid being seen from outside the high-walled compound, though there was clearly enough nonvisual evidence to support action of some kind against the compound. In the end the Obama administration took the very risky option of sending U.S. special operations forces to raid the compound and if not able to capture Bin Laden alive, to collect the body for identification.[51]

Despite losing a helicopter during the landing, the mission was successful with no U.S. casualties. The raid highlighted that unmanned options, such as cruise missiles and precision-guided bombs, lack the precision and flexibility needed for such a high-value target. Ground forces could have called off the mission if the target was not present. During the actual raid, they had to react to the original plan falling apart due to one helicopter suffering damage, and were able to recover the bodies for identification. These are things that machines cannot do presently with any degree of reliability. In the lead up to *Operation Neptune Spear*, the Obama administration had been stepping up the number of attacks carried out by armed UAV,[52] indicating that unmanned options for killing terrorists were very familiar to this president in particular.

Digital Art of the Possible

No one nation has a monopoly on the computer sciences, and the cost barriers for many areas of defense- and security-related computing are low. In some areas of civilian robotics the United States has to some extent fallen behind.[53] There is certainly competition in unmanned military systems. It should be remembered that although there has been intermittent U.S. interest in remotely piloted and unpiloted aircraft (older terminology) since World War I, it is often argued that sustained U.S. interest in UAVs was only sparked by Israeli success during the 1982 Lebanon War, where unmanned aircraft were used to both decoy air defenses into attacking nonexistent air strikes[54] and to provide critical battlefield surveillance.[55] The Israelis remain very competitive in the field of unmanned systems.[56] The People's Republic of China has certainly taken note of the recent U.S. proliferation in battlefield robotics, producing both unmanned aerial systems that are at the very least inspired by Western airframes, as well as home-grown UAS concepts.[57] More recently, subnational actors, such as the terrorist group Hezbollah, have started operating unmanned aerial systems.[58]

The capacity to outthink and outmaneuver an opponent, agility in other words, often comes up when one cannot or will not resort to overwhelming (and expensive) quantities of force. Colonel John Boyd's OODA loop is associated with this strategy, as well as the lightweight fighter concept, which led to the F-16, to capitalize on agility. Speed is an important factor in why and if computers are to be given a place in the decision to kill. The time delay between a surveillance drone seeing a target and a manned aircraft striking the target led to many targets escaping. This led to those surveillance drones being armed in an ad hoc manner with antitank missiles, embodying the entire "sensor-to-shooter" cycle in one airframe.[59]

Increasing effectiveness of weapons is no longer a crude matter of increased firepower or destructive capability; otherwise nuclear weapons would have more regular utility. Instead usable weapons are based more on precision and overwhelming speed these days. As mentioned earlier, there is a premium placed on getting ahead of the opposition in forms of human competition and conflict, and computer assistance is viewed as a key factor. However, there are those who will argue that human reaction time has become a limiting factor in these man-machine combinations. This is where fully autonomous weapons may have advantages over human-controlled weapons. The armed UAS was faster than the drone calling in manned aircraft to strike. A fully autonomous unmanned combat air system (UCAS) or other unmanned combat systems may be even faster than remote human operators at noticing and evaluating valid targets. The problem is that autonomous target perception and evaluation are only emerging military technologies in need of development. For the costs involved with developing such capability, is the increase in speed worth it? Will mistakes made by a fully autonomous weapon be viewed in a harsher light than that caused by human error, despite both mistakes being just as lethal? How much better will the autonomous weapon have to be to be fully unleashed from continuous human supervision?

The budget realities of United States in the early 21st century add to the already daunting challenge of identifying near- and short-term trends in military autonomous systems. On one hand R&D funding faces cutbacks. On the other hand, autonomous computing and unmanned warfare systems promise long-term cost savings. When the U.S. Army's Future Combat System (FCS) program for completely new manned and unmanned vehicles was cancelled in 2009, the unmanned systems survived to be independently integrated with the existing vehicles.[60] Aside from the survival of specific programs, there are also existing policy directives. Section 220 of the Floyd D. Spence National Defense Authorization Act for Fiscal Year 2001 (Public Law 106-398) wanted a third of the operational U.S. deep-strike aircraft fleet to be unmanned by 2010, and a third of operational ground-combat vehicles to be unmanned by 2015.[61] Despite the short timeframe remaining, these goals were cited in the *FY2009–2034 Unmanned System Integration Road* as part of the congressional direction that formed the basis for this Pentagon document and which still appears to be current at the time of writing.[62]

The "three Ds" used to advocate unmanned systems, tasks that are dull, dirty or dangerous, also happen to be areas where replacement of humans promises savings. Machines are already undertaking long-term surveillance missions, fuel and maintenance being the limiting factors for endurance. Analysis of the data collected could also be referred to as dull; however, the low cost of electronic reconnaissance and intelligence collection has changed the seeming monotony of scrutinizing collected intelligence to an overwhelming flood that cannot be coped with without additional personnel. A better example of cost is that of the personnel needed to maintain an effective sentry. Unmanned sensors, and smarter monitoring software, do not get tired or degrade in performance over the course of a shift.

Having people perform dirty and dangerous tasks has many costs. Financially, there are the direct costs of hazard pay and protective equipment. In the Western world if people are to be placed into harm's way then it is generally accepted that they must be properly trained and prepared to handle the risks—training and preparation that have become both more comprehensive and more expensive over time. In the end there are the human costs; people suffer the consequences of warfare, either by death, physical injury, and trauma. Although it may be cold hearted to note, the reality is that injured soldiers also have long-term costs in terms of rehabilitation and other forms of medical care that is owed to those who put themselves in harm's way on behalf of the nation.

Robotics may give options to minimize casualties, not necessarily just for the combatants removed directly from the battlefield. Remotely operated or supervised robots present the opportunity to give emotional detachment for the "man-in-the-loop," which is an opportunity to reduce the unfortunate mistakes of war brought on by stress and the pressures of surviving combat. Instead of speed being the armor, the operator would have distance as protection. An operator removed from the direct dangers of the battlefield would not have the pressures of survival forcing hasty decisions. Indeed the remotely operated vehicles or supervised autonomous robot could be in effect treated as a disposable avatar—hardware traded for a theoretical reduction in collateral damage under many circumstances. Where appropriate, the machine soldier would methodically, perhaps leisurely, select targets as it is in no rush and, as far as the operator is concerned, in no real danger. Now for some reducing the act of killing to the emotional level of answering a customer service call is disturbing and inhuman. The psychological impact on the drone's operator is also little studied at present.[63] There is, however, great appeal to the idea that a professional soldier would kill in a manner consistent with the ideals of a just war; giving the shooter time to consider their actions is one option on how to achieve this.

Then there are also the political costs of captured personnel. Not every political entity is as concerned with the proper treatment of prisoners of war as Western governments are. That there is open and continuous debate on the treatment of the small number of captured nonmilitary combatants, who are not covered by POW conventions, attests to the general consensus in the West that detainees should be treated humanely by some definitions. Historically this has not been the case elsewhere, as demonstrated by the notorious "Hanoi Hilton" and the treatment of downed Coalition aircrews in the 1991 Gulf War.

Unmanned systems, despite charges of being increasingly "gold plated" by contractors and procurement officers alike,[64] can be made financially much cheaper to own and operate. The classic example used to highlight the costs advantages of unmanned combat is the fighter pilot. Fighter pilots come from the larger pool of military aviators via an expensive selection and training process. When sent out on missions, they require life-support to function in the deadly conditions found at high altitude, G-suit to endure the effects of rapid maneuver, and as a final onboard level of protection, sit on an explosively propelled ejection seat. Depending on how far one wishes to take the accounting, a pilot's support costs also include the search and rescue resources the United States has made a duty to provide, including fleets of aircraft and helicopters, with their own crews, as well as specialist personnel, including special operations forces. Beyond this are the pension and veteran benefits due to all involved with the air combat operations.

A UCAS on the other hand could be cut down to the minimum equipment needed to reliably launch, find the target, attack, and return for recovery. If the last part, return for recovery, is omitted, this weapons system is a cruise missile. Training involving increasingly higher-resolution simulations are another avenue for savings. With virtual training there would be no need to expose the actual UCAS to the wear and tear of flight. Even if the rest of the UCAS were manufactured with the same redundancies found in manned aircraft, the lack of a pilot and associated support equipment, infrastructure, and personnel would provide considerable savings.

Autonomous Data Analysis

It is likely that the proliferation in unmanned systems for intelligence collection will create a need for advanced tools to help with analysis. Battlefield sensors are being miniaturized with the aim of providing a pervasive blanket of "smart dust." Although such surveillance coverage is unlikely to eliminate the fog of war, it will add to the friction of war by producing an overwhelming flood of data. Automating some of the pre-analysis would help with the data bottleneck, and AI will be essential to preventing such systems from becoming simple, and potentially misleading filters and information gatekeepers.

The commercial sector has produced numerous customer relations management (CRM) and data-mining tools to sift through the massive databases that consumers willing and unwillingly contribute to with every transaction. Though sometimes hard to discern, people fall into routines and patterns that are often replicated in the lives of complete strangers who fit a similar profile. Many of these events are captured in the transaction histories that banks and credit card companies keep. Other data is generated from the use of tags such as transit passes and supermarket loyalty cards.[65] Much of the personal technology that people carry, such as cell phones, actively broadcast information on a continuous basis. It is not just financial databases, and the tracking of tagged objects that are being combined, but facial-recognition technology is enabling tracking without the need for artificial markers. Surveillance cameras now use the same technology base as computer chips and have experiences a similar reduction in size and cost, accompanied by proliferation to the point of becoming another ubiquitous element of society. The now profitable industry of web searches allows for quick location of a person's "web footprint," including information a person may not realize is, or wants, in the public domain. The rapid rise of digital technology, and its implications on society, presents many policy issues for government that extend well beyond the scope of this book.

Much of this surveillance infrastructure came about as independent components for purely commercial purposes. Patterns found in the behavior of communities are clearly useful for marketing and product support. Given large enough databases, and enough computer power, higher resolution can be obtained focusing on smaller and smaller communities. Reduced cost of analysis and reduced costs of marketing, all built on ongoing advances in computing power, have led to niche or micromarketing. There are hopes and fears that this type of data collection with proper analysis will be able to someday focus advertising to the level of the individual.[66] Fraud detection would be another capability that large-scale transaction monitoring and analysis lends itself to, and already does operate at the level of individual.

Advance analysis of ambient data collected could potentially allow for preemptive identification of criminal and terrorist activity. In the aftermath of a crime, or terrorist incident, investigators have to piece together disparate clues to build up a picture of what had transpired. Investigators have been aided by computer technology, and tools that bring together the many forms of electronic monitoring occurring in the background. If criminal and terrorist acts have patterns in transactional behaviors that foreshadow the act, it then becomes in theory possible to detect these patterns before an attack or crime.[67] Developing computer systems capable of identifying trends and threats from the mass of data includes technologies such as pattern recognition and computer cognition.

As intelligence programs, the specifics of government data mining and data-analysis tools to keep up with the flood of data are kept secret; however, the interest from intelligence and military agencies is not.[68] Defense Advanced Research Projects Agency (DARPA) has publically listed many programs that would seem to be driving advances in computer pattern recognition, cognition, and autonomy. Among DARPA *Information Innovation Office* programs at the time of writing is *Deep Learning*, a program to move computer learning beyond the limited "shallow" learning available today via emulation of how biological brains work, and produce a general-purpose machine intelligence capability that can be applied to a range of applications.[69] Another program, of perhaps immediate applicability, is the *Robust Automatic Transcription of Speech* (RATS) program, which applies computer perception and cognition to language detection, identification, and analysis. This latter DARPA *Information Innovation Office* program has great resonance with the current Global War on Terror, where good translators are critical, but also, by some accounts, a hard-to-find resource.[70]

Following the death of Osama Bin Laden, a MIT paper published two years earlier entered popular media coverage of the raid—it concerned using biogeographic modeling to finding Bin Laden and appears to have been not too far off the mark. Biogeography is a branch of science normally used to model endangered species populations based on last-known sightings and habitat requirements, data points that had equivalents in the 10-year hunt for Bin Laden. Though Abbottabad was not top of the list of likely city or region he was hiding in, it was second,[71] according to the analysis done by Thomas W. Gillespie and John A. Agnew.[72] More importantly the behavioral modeling used had suggested that Bin Laden would be more likely found in an urban setting instead of in the mountainous tribal regions, as was common belief prior to the announcement that he was found and killed by U.S. forces in a major urban center. This computer model was made from the limited amount of publically available information on Bin Laden's last-known whereabouts and medical history.

Despite their common use by business, government data mining has run into controversy. First and foremost is the association of mass surveillance with oppression and tyranny, the "big brother" state of Orwellian fame. Computer technology is potentially allowing mass indiscriminate data collection and analysis to be conducted at all times on everyone participating in modern society—ubiquitous surveillance. Among the more controversial programs was DARPA's total information awareness (TIA) project, which as highlighted by its critics, may have involved mass and uncontrolled access to various types of personal data and communications.[73] Components of TIA are reputed to have continued after TIA itself was ended in the FY 2004 defense budget, meaning the legal and constitutional debate continues as well.[74]

There is also a potential for reading too much into the data. Criminals and terrorists to be successful in blending into society would be exhibiting many of the patterns of behavior of law-abiding citizens. There are concerns that in an attempt to interdict criminal or terrorist activity, innocents may be caught up. Although it is unlikely that a person's yogurt preference may lead to accusations of terrorism, there have already been incidents were datamining techniques have led to embarrassing investigative failures. One has to wonder if the online research for this and other academic works on military technology has led to placement on government watch lists (and beyond to a list of the "mostly harmless"). There is a balance between security and freedom, one that often changes with events,[75] but there are also principles, such as those that the U.S. Constitution is based on, that are sacrosanct and therefore cannot be ignored.

Air, Land, and Sea

Mobile machine intelligence faces more challenges. Apart from restrictive volume, power, and durability requirements, mobility itself presents its own perception and cognition needs. It is not surprising that UASs were the first operational examples of autonomous vehicles. Although exceptional situational awareness is prized in a military pilot, the largely empty skies provide freedom for autonomous vehicles to operate without any specific need for situational awareness. Satellite-based GPS has for all intents and purposes eliminated the ambiguities of air navigation. This lack of awareness, as mentioned earlier, has led to resistance from pilots who may have to share airspace with these machines. There is a very real danger of collision with these remotely controlled and autonomous, but unaware, aircraft.[76] Though there is a pilot, remotely piloted vehicles have their situational awareness limited by the constraints of the onboard sensors and communications bandwidth. Remotely supervised (instead of remotely piloted) implies even less situational awareness. The fact that controllers are often not pilots is a related source of resistance over expanded UAS use. With the skies becoming more crowded by UAS, there is a growing need to introduce more situational awareness to these machines.

At present most large operational UASs require an operator to be remotely controlling the vehicle during takeoffs and landing (or launch and recover in current terminology).[77] This reality is fast changing as technologies that allow for autonomous runway and taxing operations are introduced. GPS enhanced by ground beacons allows for precision navigation down to the centimeter. One of the few publicized X-37B mission objectives was to conduct a fully autonomous landing from space; it successfully performed this feat on December 3, 2010, after its first orbital flight and 244 days in orbit.[78] Of arguably greater challenge was the demonstration in 2011 of repeated "hands-free" landings on a moving U.S. Navy aircraft carrier by an F/A-18 fighter aircraft as part of the X-47B Unmanned Combat Air System Demonstrator (UCAS-D) program.[79] It should be noted that for normal F/A-18 carrier operations, the takeoff is always meant to be "hands free" with the pilot keeping away from the computer-aided flight controls until airborne.

For unmanned ground vehicles (UGVs) or unmanned ground systems (UGSs), situational awareness is paramount, not only to prevent unfortunate incidents of unmanned hit and runs, but also to avoid impassable terrain and the many other obstacles to ground movement. The DARPA grand challenges of 2004, 2005, and 2007 very publically demonstrated the initial difficulties and later fast pace of advancement with autonomous ground vehicles. From no vehicles able to successfully reach the finish line in 2004, to several being able to complete the rigorous 131.6-mile desert course the next year in

2005,[80] to several vehicles being able to navigate an urban course, with simulated traffic and parking lot to contend with, in 2007, it reflected the pace of advancement.[81] For the U.S. taxpayer, these DARPA-run competitions, with relatively small prizes of a few million dollars, also had the benefit of driving multiple lines of research among the largely self-funded competitors, including a few private and volunteer efforts. Some of this research may of course end up in personal driveways; robotics experts have been suggesting for years, that autonomous cars, with their theoretical superior adherence to the rules of the road, will cut accident rates.

UGS applications right now in development include autonomous logistic vehicles able to form convoys and deliver supplies to outposts,[82] and smaller multilegged machines to follow dismounted infantry through rough terrain.[83] Short range, remotely controlled, robots, practically the same as used by police bomb disposal, became a common piece of equipment for explosive ordnance disposal (EOD) units operating in Iraq and Afghanistan. On a larger scale would be equipping future long-range autonomous UGSs with mine-sweeping equipment, relieving sappers and EOD of a task that can be both very time consuming and monotonous, arguably fitting the dull category, but inescapably dangerous. Mines and improvised explosive devices (IEDs) are a prominent danger for supply convoys, meaning UGS technologies would not only be reducing the dangers to drivers, but also increasing the odds of supplies getting through.

Armed UGSs, variants of existing EOD robots, exist but there currently seems to be a reluctance to actually use them in combat.[84] These are still remotely controlled machines, and many share technology with remote weapon station (RWS) turrets mounted on some armored vehicles. Advantages given for an RWS over a traditional manned turret include the RWS control station taking up less room inside the vehicle than a traditional turret basket, and increased protection by having the gunner fully inside the armored vehicle. Drawbacks include less situational awareness by virtue of the outside world being condensed into a video stream, and lack of access to the weapon if it jams. The armed (remotely controlled) robot in this sense is the same as the RWS except that the operator is given much more separation from the weapon, which conceptually gives more protection. The increased remoteness, however, aggravates the drawbacks, and introduces the problem of a vulnerable control signal. With fears of cyber warfare, both the combat robot and RWS face similar threats from hacking, though the wireless connection to the robot would still seem more vulnerable.

Concerns over the enemy taking over control of a UGS highlights the overall problem of maintaining a data link. Without an autonomous capability, a drone that is cut off from its controller is just as dead as if it had been blown up. Fears over the ability for peer competitors, and others, to be able to override the control of drone operator is quite clearly a worse scenario. There have already been reports of the data stream from the UAS being intercepted.[85] The need for a secure data link makes remotely controlled unmanned system is more vulnerable than a living breathing soldier. A soldier can not only function, but also still fight, without a digital umbilical to HQ. The imprecise nature of psychology makes it much harder, arguably impossible, to break in and usurp leadership of the enemy's forces with any degree of reliability. A simple two-way data flow on the other hand requires just hacking encrypted communications and control software. A UGS operating in a busy urban environment is perhaps more susceptible. The urban jungle does not present a clean radio-frequency environment, with many transmitters competing for reception. Hostile electronic countermeasures, even without attempts to take over, add to the problem of communicating with the UGS.

The maritime environment presents its own challenges. It is better to consider maritime operations as two separate battle spaces: surface and underwater. Unmanned surface vehicles (USVs) can make use of GPS navigation, allowing them to use similar navigation techniques as UASs and UGVs. Maritime security, protecting shipping in harbors, and conducting short-range patrols are tasks that USVs are being marketed for. The unmanned undersea vehicle (UUV) has a more difficult navigational task, though one that is familiar to submariners: navigation without regular access to the outside world. The tethered approach used by remotely operated submersibles used in research such as deep-sea exploration and for underwater industry would limit military applications to only those that could be near a mother ship.[86] The difficulty in maintaining constant communications with a UUV would seem to make some degree of autonomous operation a necessity. Along with computer autonomy, accurate low-cost inertial navigation and onboard sonar technology have also benefited from the information-technology-based miniaturization. Potential applications are broadly similar to that in other environments and include minesweeping and reconnaissance of landing areas.

A potential technological choice in unmanned military systems is that between relatively large self-contained machines, or larger groups of smaller, individually less capable, machines acting together toward a common goal, but without any direct coordination or centralized command, a swarm. In nature, very small creatures, often of limited intelligence, self-assemble into groups (swarms, flocks, packs, etc.) to carry out tasks such as migration and hunting. Although it may be impossible to create human-level intelligence (and therefore impossible to achieve the technological singularity), it may be possible to replicate the functional intelligence of an ant, locust, bird, fish, or a wolf—all types of types of creatures that have examples of traversing great distances together and, for the military, even provide examples where groups hunt together. The swarm concept leverages off of the still ongoing trend of computers and sensors shrinking in size and price, and combines it with the study of successful examples from nature.

The military robots swarm already has a bit of a precursor in the form of the CBU-97 sensor fused weapon (SFW). The SFW is a cluster bomb containing 40 hockey-puck-sized submunitions, called "skeets," each equipped with multiple sensors (infrared to detect engines, laser contour sensor to detect the shape of a vehicle), and an explosively formed penetrator warhead. Unlike robots in the swarm concept proper, skeets are spread out by a complex dispensing mechanism: air bags eject a pattern of 10 parachute-stabilized dispensers from the CBU-97; these dispensers then use rockets to spin the dispensers and fling from each four skeets each over a targeted area. During the brief operational life of a skeet it will on its own find a target and determine the best position during its intentionally wobbling decent to attack; any that fail to engage self-destruct.[87] A single SFW can attack multiple targets in a 121,400-square-meter (30 acres) area.[88] The SFW is used to attack multiple easy-to-see targets out in the open, a convoy for instance, that the pilot would need to find before releasing the cluster bomb—a cluster bomb in accordance with precision-warfare paradigms.

A robot swarm on the other hand would be using the many sensors distributed among the swarm to seek out targets, and then coordinate an attack against found targets. Being self-propelled, a robot swarm would be able to survey hundreds if not thousands of square kilometers. A horde of small sensors, with smaller fields of view can, for some elusive targets, be more efficient than a single large sensor scanning a large area. On finding a target, the swarm would assemble to cooperatively attack. After using only as much force as needed to destroy the target, the swarm would disperse again to find more targets. The swarm concept is just one too many that leverage off of expected near-term advances in robotics to provide first persistent surveillance, and later persistent attack over entire regions. Specifically, the swarm concept utilizes not only computer autonomy, but also the ability for a group of machines to self-organize for the different needs of general surveillance and carrying out an attack. This is not just trust in the machine, but trust in the aggregate of many machines. As with many network-enabled concepts, there is room for human supervision. The degree of human supervision and the centralization of command and control are indicative of whether a specific group of robots is operating as a swarm, or tied to a controlling "mother ship"—with ultimately trust in the technology deciding on these factors.[89]

The swarm concept mirrors how computers are being networked together today to handle tasks that a single monolithic computer would have difficulty with. Parallel computers have been put together from used desktop computers and gaming consoles to provide low-cost supercomputer-like capabilities. Parallel processing and today's large bandwidth networks are allowing the creation of these supercomputers on an ad hoc basis, with elements spread across a large geographic region. Computer criminals already use similar techniques to overwhelm targeted servers with a distributed denial of service (DDOS) attack. Using thousands of hijacked personal computers, these machines are remotely commanded to flood the targeted server with network traffic. Hijacked computers, called zombie computers, formed into remotely control networks, called botnets, are hired out for criminal enterprises and spam e-mail purposes. It would seem that reliable covert RC over thousands of remote computer devices is not an insurmountable requirement.

From Advanced Smart Weapon to Semi-Autonomy and Beyond

Many defense academics have been noting the trend of increasing area of battlefield allocated per combatant[90] or, similarly, the increasing range of weapons with time. If shear destructive power is being avoided, then increased accuracy is required to compensate—this is the raison d'être for precision-guided weapons. Ranges and battle spaces have increased in size to the point where it is difficult for the human mind to handle. This has created a need for various command, control, communications, computers, intelligence, surveillance, and reconnaissance (C4ISR) tools, with increasing amounts of automation to assist the warfighter. In other words, the current trends and needs in conventional warfare are providing plenty of opportunities that can be filled by AI advances.

Now there is reluctance for both real operational and acceptance reasons to remove the "man-in-the-loop"; so perhaps it is better to say that semi-autonomy is the direction that weapons and platforms are taking, in the near term at least. Indeed with the difficulties involved with developing a robust form of AI able to cope with the real world, it would seem that people will not be cut completely from the loop for some time to come. However, computer power is increasing at an exponential rate, and systems are becoming increasingly capable of at least mimicking many decision-making processes. Increased computer power is aiding research in other areas of science, including the nature of consciousness and sentience. These advances are converging, with perhaps long-term implications on just how much trust future leadership and warfighters are willing to place on machines that can tell when to kill.

Notes

1. Philip Aston, United Nations Human Rights Council Report "Report of the Special Rapporteur on Extrajudicial, Summary or Arbitrary Executions, Philip Alston," May 28, 2010, http://www2.ohchr.org/english/bodies/hrcouncil/docs/14session/A.HRC.14.24.Add6.pdf.

2. United Kingdom, Department for Business Innovation & Skills, *A.I. Law: Ethical and Legal Dimensions of Artificial Intelligence*, October 5, 2011, http://www.sigmascan.org/Live/Issue/ViewIssue/485/1/a-i-law-ethical-and-legal-dimensions-of-artificial-intelligence/.

3. P. W. Singer, *Wired for War* (New York: The Penguin Press, 2009), 337–38.

4. See Glossary.

5. Douglas Mulhal, *Our Molecular Future—How Nanotechnology, Robotics, Genetics, and Artificial Intelligence Will Transform Our World* (Amherst: Prometheus Books, 2002), 61.

6. Real numbers can have an infinite number of digits on both sides of the decimal places. For instance, in real numbers between 0 and 1 there are an infinite number of divisions. The binary system is an integer-only number system—whole numbers only.

7. This claim was made early in at least one introduction to a digital logic course for engineering students.

8. Intel, "Moore's Law and Intel Innovation," http://www.intel.com/about/companyinfo/museum/exhibits/moore.htm.

9. Michael Kanellos, "Moore's Law to Roll On for Another Decade," *CNET*, February 10, 2003, http://news.cnet.com/2100–1001–984051.html.

10. Ibid.

11. Ray Kurzweil, "How My Predictions Are Faring," October 2010, http://www.kurzweilai.net/predictions/download.php.

12. "The Secret to Google's Success," *Bloomberg Businessweek*, March 6, 2006, http://www.businessweek.com/magazine/content/06_10/b3974071.htm.

13. Don Reisinger, "Finland Makes 1Mb Broadband Access a Legal Right," *CNET*, October 14, 2009, http://news.cnet.com/8301–17939_109–10374831–2.html.

14. P. W. Singer, *Wired for War* (New York: The Penguin Press, 2009), 189.

15. Ibid., 194.

16. United States Internet Crime Complaint Center, "2010 Internet Crime Report," http://www.ic3.gov/media/annualreports.aspx.

17. Larry Greenemeier, "Seeking Address: Why Cyber Attacks Are So Difficult to Trace Back to Hackers," *Scientific American*, June 11, 2011, http://www.scientificamerican.com/article.cfm?id=tracking-cyber-hackers&WT.mc_id=SA_Twitter_sciam.

18. Department of Defense, *Military and Security Developments Involving the People's Republic of China 2010*, http://www.defense.gov/pubs/pdfs/2010_CMPR_Final.pdf.

19. Department of Defense, *Department of Defense Strategy for Operating in Cyberspace*, July 2011, http://www.defense.gov/news/d20110714cyber.pdf.

20. British Broadcasting Corporation, "Stuxnet Worm Hits Iran Nuclear Plant Staff Computers," September 26, 2010, http://www.bbc.co.uk/news/world-middle-east-11414483.

21. Sergey Golovanov, "TDL4 Starts Using 0-Day Vulnerability!" *Securelist*, http://www.securelist.com/en/blog/337/TDL4_Starts_Using_0_Day_Vulner ability.

22. Fahmida Y. Rashid, "Marine General Calls for Stronger Offense in U.S. Cyber-Security Strategy," *eWeek.com*, July 15, 2011, http://www.eweek.com/c/a/IT-Infrastructure/Marine-General-Calls-for-Stronger-Offense-in-US-CyberSecurity-Strategy-192629/.

23. John Palfrey, "Middle East Conflict and an Internet Tipping Point," *Technology Review*, February 25, 2011, http://www.technologyreview.com/web/32437/?mod=chthumb.

24. Michael Holden, "Make Cupcakes, Not Bombs," *Reuters*, June 3, 2011, http://uk.reuters.com/article/2011/06/03/uk-britain-mi6-hackers-idUKLNE 75203220110603.

25. Rodney Brooks, *Flesh and Machines: How Robots Will Change Us* (New York: Pantheon Books, 2002), 43.

26. The name *Southern Cross II* is a tribute to the historic 1928 United States to Australia flight of the original *Southern Cross*.

27. Dr. Jim Young, Air Force Flight Test Center History Office, "Milestones in Aerospace History at Edwards AFB," June 2011, http://www.af.mil/shared/media/document/AFD-080123–063.pdf.

28. A tribute to both Charles Lindbergh's *Spirit of Saint Louis*, and the farm where it was tested.

29. Fédération Aéronautique Internationale, "Aeromodelling and Spacemodelling records," http://www.fai.org/ciam-records.

30. "Trans Atlantic Model," http://tam.plannet21.com/.

31. United States, "Iris Recognition," August 2007, http://www.biometrics.gov/Documents/irisrec.pdf.

32. Mark Milian, "Facebook Lets Users Opt out of Facial Recognition," *CNN*, June 9, 2011, http://www.cnn.com/2011/TECH/social.media/06/07/facebook.facial.recognition/index.html?iref=allsearch.

33. Mark Milian, "Google Taking More Cautious Stance on Privacy," *CNN*, June 1, 2011, http://www.cnn.com/2011/TECH/web/05/31/google.schmidt/index.html.

34. Joseph A. Bernstein, "Seeing Crime Before it Happens," Discovery, January 23, 2012, http://discovermagazine.com/2011/dec/02-big-idea-seeing-crime-before-it-happens.

35. The sensor fused weapon program uses multiple sensors to identify viable vehicle size targets.

36. Cathryn M. Delude, "Computer Model Mimics Neural Processes in Object Recognition," MIT press release, February 23, 2007, http://web.mit.edu/press/2007/surveillance.html.

37. Ibid.

38. Sensor fused weapon again.

39. P. W. Singer, *Wired for War* (New York: The Penguin Press, 2009), 126.

40. David Hoffman, "I Had a Funny Feeling in My Gut," *Washington Post*, February 10, 1999, http://www.washingtonpost.com/wp-srv/inatl/longterm/coldwar/shatter021099b.htm.

41. Russian Federation. "On Presentation of the World Citizens Award to Stanislavpetrov," January 19, 2006, http://www.un.int/russia/other/060119eprel.pdf.

42. Whether the sensors were in error is a different matter.

43. Nicholas Thompson, "Inside the Apocalyptic Soviet Doomsday Machine," *Wired*, Issue 17.10, September 21, 2009, http://www.wired.com/politics/security/magazine/17–10/mf_deadhand.

44. Philip Aston, United Nations Human Rights Council Report "Report of the Special Rapporteur on Extrajudicial, Summary or Arbitrary Executions, Philip Alston," May 28, 2010, http://www2.ohchr.org/english/bodies/hrcouncil/docs/14session/A.HRC.14.24.Add6.pdf.

45. Ibid.

46. Philip Aston, United Nations Human Rights Council Report, "Report of the Special Rapporteur on extrajudicial, summary or arbitrary executions, Philip Alston," May 28, 2010, http://www2.ohchr.org/english/bodies/hrcouncil/docs/14session/A.HRC.14.24.Add6.pdf.

47. *Convention on the Prohibition of the Use, Stockpiling, Production and Transfer of Anti-Personnel Mines and on their Destruction*, September 18, 1997, http://www.icrc.org/ihl.nsf/FULL/580?OpenDocument.

48. Amnesty International, *'Less Than Lethal'? The Use of Stun Weapons in US Law Enforcement*, December 16, 2008, http://www.amnesty.org/en/library/info/AMR51/010/2008/en.

49. Nat Hentoff, CATO Institute, "Few Batting Eyes at Obama's Deadly Drone Policy," *Cato.org*, July 29, 2009, http://www.cato.org/pub_display.php?pub_id=12012.

50. Mark Mazzetti, Helene Cooper, and Peter Baker, *New York Times*, May 2, 2011, "Behind the Hunt for Bin Laden," http://www.nytimes.com/2011/05/03/world/asia/03intel.html?_r=1.

51. CBS, "Obama on Bin Laden: The Full "60 Minutes" Interview," May 8, 2011, http://www.cbsnews.com/8301–504803_162–20060530–10391709.html.

52. Paul McLeary, Sharon Weinberger, and Angus Batey, "Drone Impact on Pace of War Draws Scrutiny," *Aviation Week*, July 8, 2011, http://www.aviationweek.com/aw/generic/story_generic.jsp?channel=dti&id=news/dti/2011/07/01/DT_07_01_2011_p40–337605.xml&headline=Drone%20Impact%20On%20Pace%20Of%20War%20Draws%20Scrutiny.

53. Larry Greenemeier, "National Robotics Week to Highlight the Past, Present and Future of Robot Research," *Scientific American*, February 9, 2010, http://www.scientificamerican.com/blog/post.cfm?id=national-robotics-week-to-highlight-2010–02–09.

54. P. W. Singer, *Wired for War* (New York: The Penguin Press, 2009), 56.

55. Charles Levinson, "Israeli Robots Remake Battlefield," *Wall Street Journal*, http://online.wsj.com/article/SB126325146524725387.html#MARK.

56. Ibid.

57. Graham Warwick, "China Targets UAS as Growth Sector," *Aviation Week*, May 5, 2011, http://www.aviationweek.com/aw/generic/story.jsp?id=news/awst/2011/04/25/AW_04_25_2011_p62–312195.xml&channel=defense.

58. Peter La Franchi, "Iranian-Made Ababil-T Hezbollah UAV Shot Down by Israeli Fighter in Lebanon Crisis," *Flight International*, August 15, 2006, http://www.flightglobal.com/articles/2006/08/15/208400/iranian-made-ababil-t-hezbollah-uav-shot-down-by-israeli-fighter-in-lebanon.html.

59. David Eshel, "Technology Shortens the Kill Chain in Urban Combat," *Aviation Week—Defense Technology International*, March 28, 2008, http://www.aviationweek.com/aw/generic/story_generic.jsp?channel=dti&id=news/DTIKILL.xml&headline=Technology%20Shortens%20the%20Kill%20Chain%20in%20Urban%20Combat.

60. Kris Osborn, "FCS Is Dead; Programs Live On," *Defense News*, May 18, 2009, http://www.defensenews.com/story.php?i=4094484.

61. *National Defense Authorization Fiscal Year 2001*, http://www.dod.mil/dodgc/olc/docs/2001NDAA.pdf.

62. Department of Defense, *FY2009–2034 Unmanned Systems Integration Roadmap*, 2009, http://www.acq.osd.mil/psa/docs/UMSIntegratedRoadmap2009.pdf.

63. Paul McLeary, Sharon Weinberger, and Angus Batey, "Drone Impact on Pace of War Draws Scrutiny," *Aviation Week*, July 8, 2011, http://www.aviationweek.com/aw/generic/story_generic.jsp?channel=dti&id=news/dti/2011/07/01/DT_07_01_2011_p40–337605.xml&headline=Drone%20Impact%20On%20Pace%20Of%20War%20Draws%20Scrutiny.

64. P. W. Singer, *Wired for War* (New York: The Penguin Press, 2009), 256–59.

65. Adam L. Penenberg, "The Surveillance Society," *Wired*, Issue 9.12, December 2001, http://www.wired.com/wired/archive/9.12/surveillance.html.

66. Clay Dillow, "IBM's Digital Billboard Displays Individualized Ads by Reading the RFID Data in Your Wallet," *Popular Science*, August 2, 2010, http://www.popsci.com/technology/article/2010–08/ibms-new-digital-billboard-tailors-individual-ads-rfid-data-your-credit-card.

67. John Markoff, "Taking Spying to Higher Level, Agencies Look for More Ways to Mine Data," *The New York Times*, February 25, 2006, http://www.nytimes.com/2006/02/25/technology/25data.html?ref=johnmarkoff&pagewanted=print.

68. Ibid.

69. DARPA, "Deep Learning," http://www.darpa.mil/Our_Work/I2O/Programs/Deep_Learning.aspx.

70. Jason Straziuso, "US Companies Send Translators to Afghanistan Who Are Old, Out Of Shape, Unprepared For Combat," *Huffington Post*, http://www.huffingtonpost.com/2009/07/22/us-companies-send-transla_n_243046.html.

71. http://www.bbc.co.uk/news/world-13275104.

72. Thomas W. Gillespie and John A. Agnew, "Finding Osama Bin Laden: An Application of Biogeographic Theories and Satellite Imagery," *MIT International Review*, February 17, 2009, http://web.mit.edu/mitir/2009/online/finding-bin-laden.pdf.

73. Mark Williams, "The Total Information Awareness Project Lives on," *Technology Review*, April 26, 2006, http://www.technologyreview.com/communications/16741/.

74. Ibid.

75. Adam L. Penenberg, "The Surveillance Society," *Wired*, Issue 9.12, December 2001, http://www.wired.com/wired/archive/9.12/surveillance.html.

76. Peter La Franchi, "Animation: Near Misses between UAVs and Airliners Prompt NATO Low-Level Rules Review," *Flight International*, March 14, 2006, http://www.flightglobal.com/articles/2006/03/14/205379/animation-near-misses-between-uavs-and-airliners-prompt-nato-low-level-rules.html.

77. John A. Tirpak, "The RPA Boom," *Airforce Magazine*, August 2010, http://www.airforce-magazine.com/MagazineArchive/Documents/2010/August%20 2010/0810RPA.pdf.

78. Scott Fontaine, "X-37B Test Mission Called Big Accomplishment," *Defense News*, December 6, 2010, http://defensenews.com/story.php?i=5176376&c= AME&s=TOP.

79. Graham Warwick, "F/A-18 Shows UCAS-D Can Land on Carrier," *Aviation Week*, July 8, 2011, http://www.aviationweek.com/aw/generic/story_channel. jsp?channel=defense&id=news/asd/2011/07/08/05.xml&headline=F/A-18%20 Shows%20UCAS-D%20Can%20Land%20On%20Carrier.

80. DARPA, "A Huge Leap Forward in Robotics R&D: $2 Million Cash Prize Awarded to Stanford's 'Stanley' as Five Autonomous Ground Vehicles Complete DARPA Grand Challenge Course," News Release, October 9, 2005, http://archive. darpa.mil/grandchallenge05/gcorg/downloads/GC05%20Winner.pdf.

81. DARPA, "Urban Challenge," http://archive.darpa.mil/grandchallenge/ index.asp

82. General Dynamics, http://www.gdrs.com/robotics/programs/program.asp? UniqueID=12.

83. Marc Raibert, Kevin Blankespoor, Gabriel Nelson, et al. "BigDog, the Rough-Terrain Quaduped Robot," http://www.bostondynamics.com/img/BigDog_IFAC_ Apr-8–2008.pdf.

84. Erik Sofge, "The Inside Story of the SWORDS Armed Robot "Pullout" in Iraq: Update," *Popular Mechanics*, October 1, 2009, http://www.popularmechanics. com/technology/gadgets/4258963.

85. Siobhan Gorman, Yochi J. Dreazen, and August Cole, "Insurgents Hack U.S. Drones," *Wall Street Journal*, December 17, 2009, http://online.wsj.com/article/ SB126102247889095011.html.

86. Underwater salvage and communications line tapping are important military and intelligence applications where tethered remote systems have been used. See: Sherry Sontag, Christopher Drew, and Annette Lawrence Drew, *Blind Man's Bluff: The Untold Story of American Submarine Espionage* (New York: Public Affairs, 1998).

87. Textron Systems. "Sensor Fuzed Weapon," http://www.textrondefense.com/ assets/pdfs/datasheets/sfw_datasheet.pdf.

88. Ibid.

89. P. W. Singer, *Wired for War* (New York: The Penguin Press, 2009), 229–36.

90. Doug Beason. *The E-Bomb* (Cambridge, Massachusetts: Da Capo Press, 2005), 33.

CHAPTER 5

Nanotechnology

Nanotechnology, engineering on the nanometer (one-billionth of a meter) scale is often touted as the next "big thing" for society at large, and has already a host of existing military applications. The prefix "nano" is being used to promote, hype in many cases, many current and future products. There is controversy over many long-term aspects of nanotechnology, including debate over the feasibility of "assemblers," multipurpose nanoscale devices capable of producing anything it has feed material for, including copies of itself. Speculation on nanotechnology ranges from technological singularity utopias to end-of-the-world scenarios. Predicted security and defense applications are no less ambitious, or controversial.

In general, nanotechnology is often heralded as a disruptive technology for the business world, specifically for investors looking to ride the next major wave in the economy; however, destabilization of existing economic norms is also potential fuel for conflict. Nanotechnology prophecies of eliminating material disputes in the very long term are beyond the scope of this discussion. Although it is likely nanotechnology will change society, the outcomes may not be so optimistic. Like any powerful tool, the biggest threat from nanotechnology is that someone else may use it to change the world to their version of paradise on earth.

Science and Engineering at the Nanometer Scale

The prefix "nano" refers to one-billionth—one nanometer being one-billionth of a meter. In comparison, the useful scale for cells in the human body is the micrometer (micron) scale—one-millionth of a meter. Nanotechnology in general involves products and controlled processes on the scale of cellular organelle (the "machinery" or "organs" inside a cell) and smaller. The U.S. government's National Nanotechnology Initiative (NNI) uses the following definition of nanotechnology:

> Nanotechnology is the understanding and control of matter at dimensions between approximately 1 and 100 nanometers, where unique phenomena enable novel applications. Encompassing nanoscale science, engineering, and technology, nanotechnology involves imaging, measuring, modeling, and manipulating matter at this length scale.
>
> A nanometer is one-billionth of a meter. A sheet of paper is about 100,000 nanometers thick; a single gold atom is about a third of a nanometer in diameter. Dimensions between approximately 1 and 100 nanometers are known as the nanoscale. Unusual physical, chemical, and biological properties can emerge in materials at the nanoscale. These properties may differ in important ways from the properties of bulk materials and single atoms or molecules.[1]

The phrase "unique phenomena," refers to natural forces and interactions that manifest, or become a consideration, only at the very small scale. The academic notions that the atoms and molecules are in constant motion (relatively immobile vibrating of atoms in solids, much more random and mobile Brownian motion in gases and liquids) are major considerations for engineering and manufacturing at the nanoscale. In discussions on nanotechnology the effects of viscosity and surface tension become more pronounced as the object or system approaches the size of living cells and indeed approaches the size of the molecules that make up the fluid. The effects of scale and surface tension allow insects to walk on water whereas larger creatures must swim. These are all existing physical phenomena; only that common everyday experience at the large scale does not give us an intuitive appreciation of their effects at the very small scale.[2]

An example of unintuitive nanoscale effects is the replication of gecko setae, flexible nanoscale hairs covering a gecko's feet, which allow it to cling to surfaces without adhesive.[3] Both natural and artificial setae use the van der Waals force[4] to generate attraction between surfaces. The van der Waals force only becomes usable with extremely close contact between the molecules of different objects. Setae as nanoscale structures are able to press into rough textures of most surfaces, allowing a useful amount of van der Waals force to manifest.[5] Engagement of the setae is directional, meaning a gecko can stick or unstick a foot with only a subtle movement of the limb. Both the strength and controllability of this adhesive-free stickiness has been replicated in the form of artificial setae formed into mass nanoscale arrays. Gecko-like climbing and clinging abilities have surveillance and military uses.[6] Manufacturing quantities of artificial setae is, however, a barrier toward real-world use.

Although the origins of nanotechnology are often attributed to Richard Feynman's 1959 lecture, "There's Plenty of Room at the Bottom,"[7] there are examples of regular use of nanoparticles from well before this. A commonly given example of nanotechnology from the period before the concept of nanotechnology became known is stained glass. Though the science was not recognized for centuries, some colors of stained glass are the result of correctly sized nanoparticles of gold and other materials forming in the glass.[8] This phenomenon, structural color, is the result of the shape of the particle scattering and reflecting specific wavelengths of light, differing from pigments that absorb and reflect specific wavelengths of light. The visible portion of the spectrum has wavelengths in the hundreds of nanometers. Today's understanding of how light interacts with matter is being developed into increased control over light and other forms of electromagnetic (EM) radiation, the principles behind quantum dots and metamaterials.

There is debate over the definition and goals of the nanotechnology field centering on the concept of bottom-up, guided atom-by-atom construction of structures and devices. In policy, and in marketing, nanotechnology simply refers to products that are or work at less than 100 nanometers and does not make any distinction between how it is produced. Nanotechnology, as popularized by K. Eric Drexler in works such as the 1986 book, *Engines of Creation*, and later in the 1993 book, *Unbounding the Future: The Nanotechnology Revolution*, is very specific in referring to nanotechnology as being atomic-level construction, building up products via what are termed "molecular assemblers," or simply, "assemblers." Some have labeled this definition for nanotechnology as the "radical vision"[9] or "futuristic nanotechnology."[10] Not only is it a break from established industrial models, but also promises to revolutionize the world by changing economic paradigms. If an atomic- and molecular-scale assembler were found to be possible then it could build any useful object, whether it is food, shelter, weapons, or copies of the assembler, efficiently with minimal feed materials, energy, and time. Thus far the molecular assembler remains theoretical; there is debate over whether such devices are practical.

Before discussing the nanotechnology assembler, it is useful to examine how nanoscale products are manufactured today—top-down processes. Top-down fabrication processes have physically large systems producing nanoscale products. Semiconductor computer chips, although not specifically labeled as nanotechnology, are a good example of the top-down approach. Mass production of computer chips is done in expensive fabrication plants where thin layers of material are alternatively coated on and removed to produce the necessary features of a chip. The most common technique used to produce computer chips is photolithography—the use of light and shadows to selectively activate light-reactive materials controlling whether they stay or are removed during the steps of production. With 32 nanometer features already common, and 22 nanometer products entering production as this is being written, the technology has moved far beyond the use of visible light—the color violet at around 400 nanometers is the smallest wavelength perceptible.[11] The equipment used at each stage of manufacture is many times larger than the computer chips being manufactured, and indeed the "image" used in photolithography to define the chips is many times larger than the image cast on the silicon. The large investment found in retooling a semiconductor fabrication plant for every new generation of computer chip is a major problem when product obsolescence is measured in mere months.

Carbon nanotube (CNT), a product definitely associated with nanotechnology, is presently produced via top-down industrial processes. CNTs are particularly long chains of just carbon atoms that exhibit strength and conductivity properties which are of interest to a wide range of disciplines. Specifics of how the carbon atoms are arranged define the different mechanical and electrical properties of the different forms of carbon. Like CNT, diamonds and graphite are made up of just carbon arranged under specific conditions. Miniscule quantities of CNT form naturally, but in the quantities needed even for experimentation, it is a manufactured commodity. Carbon-bearing feed materials (organic molecules in other words) and catalysts are exposed to high temperatures and other conditions needed to first breakdown the organic molecules and then coax the carbon atoms to bond with other carbon atoms correctly to form the large molecules of the specific type of CNT being produced.[12] Noncarbon elements from the feed material as well as the catalyst represent waste products that may be processed to recover useful elements or discarded.

Though only theorized, a bottom-up process would involve assembly of components atom by atom and molecule by molecule. For several decades now, technologies such as the scanning tunneling microscope (STM) and the atomic force microscope (AFM) have been pushing atoms and molecules around, one at a time. As most manufactured goods are composed of trillions of atoms, single atom-by-atom placement is of little use outside of very specialized research and media events.[13] For anything useful to be mass produced, the individual placement of atoms and molecules would have to be multiplied into parallel-assembly processes where many atoms and molecules are deliberately moved and placed simultaneously. Parallel assembly leads to the nanotechnology assembler, independent nanoscale machines able to arrange atoms and molecules one at a time. Parallel work by thousands, if not millions of assemblers, is necessary for this concept to actually produce a high-technology product before it becomes obsolete.

One immediate problem is how start off with enough assemblers in the first place. As in all finely crafted machines, and it does not get any finer than the atomic level, the first assembler is expected to be a very expensive item. Discussions on assemblers led to the concepts of nanotechnology assemblers building more assemblers and exponential growth—in the time that it takes an assembler to build a copy of itself, a generation, the number of assemblers double. Although slow to start, the exponential growth of assemblers is envisioned to quickly take care of the numbers problem. Handling seemingly insurmountable tasks with self-replicating machines and systems is not unique to nanotechnology; mathematician John Von Neumann is generally credited with the idea of factories, machines, and systems (not necessarily at the nanoscale) building copies of themselves in the 1940s, leading these systems in general to be referred to as Von Neumann machines. These have appeared in science fiction, usually as a malevolent all-consuming force, though in Arthur C. Clarke's *2010: Odyssey Two* it seems more a matter of the self-replicating monoliths being too busy to deliver more than a cryptic safety notice about their use of Jupiter as a construction zone. Recent alarm over nanotechnology assemblers running amok and consuming the world, the "gray goo" end-of-the-world scenario, is therefore nothing new. The term Von Neumann machine is even noted in the endnotes of Clarke's novel along with contemporary NASA interest in such systems to quickly and cheaply perform large-scale tasks.[14] From the perspective of the theoretical nanotechnology assembler, fabricating something sized for human use is at first an impossibly large task, akin to the rapid industrialization of the moon that NASA was publically discussing in the 1980s. Like the nanotechnology assembler, the industrialization of the moon is not physically impossible, but faces great engineering challenges and even greater economic barriers.

Although autonomous self-assembling machines or systems would seem to be an overused science fiction cliché, proponents are quick to cite that their inspiration is nature itself. All multicell organisms, such as people, start off as a single cell. Chicken eggs are essentially a large single cell—when fertilized, and given the correct conditions, this one cell divides and differentiates until it forms a chicken, which can go off and produce more eggs, each of which can develop into another chicken. Life itself is by definition self-assembling and capable of exponential growth; given time, food, and security, all forms of life will seek to fill a given environment.

Many of the techniques suggested for use in a nanotechnology assembler, such as DNA-like computer tape, are borrowed directly from biology. Drexler speculated on direct mechanical and chemical methods of data input and storage at the nanoscale, sidestepping the problem of building electronic subsystems to fit inside the already minimalistic assemblers. In several respects, this radical vision is much less ambitious than nature in that it is not aiming to produce initially anything as complex as a living cell. This is certainly true for pools of assemblers producing passive structural materials in bulk. The assembler form of nanotechnology promoted by Drexler and others faces many practical engineering difficulties but as a concept is not ruled out as being impossible due to the laws of science; if it were so then life would be impossible.[15]

Another aspect of the molecular assembly is that generally these are described as low-energy processes. Low-energy processes again mimic life, where slow chemical reactions generate all the organic (carbon containing) molecules, and energy needed to form first amino acids, then proteins, cells, and finally multicellular organisms. Most proponents of nanotechnology assembly see it as an efficient method to essentially "grow" all the products needed in future, and bring into reality projects that previously were too expensive to contemplate, such as armoring buildings against a wider range of natural disasters and manmade attacks.[16] The low energy requirements for production of almost any product do of course have security implications if this capability first were to exist and, second, were to proliferate. In its ultimate form, the molecular assembler would minimize waste products by virtue of only using the molecules it needed and, unlike conventional top-down processes, not by starting with a bulk of material and carving it down to a finished product. That said, depending on what feed materials are needed there is still potential for industrial waste; if one is trying to produce atomically perfect steel, one has to still start off with raw ores and what is left after refinement is still a slag heap of some kind. Low energy simply means it may be viable to do more processing of the waste materials, but only if one chooses to.

It is also important to note that the below the scale of designer molecules is the world of high-energy physics and nuclear technology. Atoms are composed of protons, neutrons, and electrons. Chemical reactions occur with exchanges of electrons but do not affect atomic nuclei, which are cores of protons and neutrons. An element is defined by the number of protons in its atoms. Isotopes, or variants of an element, are defined by the number of neutrons. As an example, all atoms of the element hydrogen have one proton in the nucleus. Deuterium, an isotope of hydrogen, has a nucleus made up of one proton and one neutron. The syntheses of atoms, the transmutation of an atom of one element into another, are nuclear reactions.

Natural nuclear decay has large, relatively unstable, isotopes spontaneously shedding protons and neutrons to achieve lower energy (more stable) configurations. As protons and neutrons are involved, atomic decay transmutes atoms of one element into those of another. These released high-energy particles also constitute forms of ionizing radiation. Alpha decay is the release of an alpha particle, composed of two protons and two neutrons, which is otherwise itself an atom of helium, and specifically the isotope helium-4. Alpha particles cannot penetrate far, but are dangerous if a source of this radiation finds its way inside the body. This process, although natural and spontaneous, does release energy that may be harnessed as is done in radioisotope thermoelectric generators (RTGs). At the time of writing, over 30 years after launch, the RTGs on the deep space *Voyager* probes were still operating. Induced fission nuclear reactions use high-energy particles to split a large atom. Depending on the atoms involved, the fission reaction may be self-sustaining through the release of high-energy particles to continue the process. Controlled, this is a fission nuclear reactor; uncontrolled, this is the chain reaction of an atomic warhead.

Of even greater energy is nuclear fusion—the process for building up atoms. Nuclear fusion takes atoms of lighter elements and forces them together to form heavier elements. In large particle accelerators, atoms can be smashed together to build up heavier elements, including short-lived unstable elements not found in nature. In thermonuclear warheads a fission explosive is used to initiate fusion in particular isotopes of hydrogen and lithium. Efficient controlled fusion, able to generate more usable energy than it consumes, has been continuously promised as being only a few decades away for the last few decades. In nature, the fusion of different elements describes the life cycle of stars—hydrogen, the simplest of atoms, fuses into helium initially. At the end of a star's life, fusion involves progressively heavier elements, thus producing all elements found in the natural universe.

Below the scale of neutrons, protons, and electrons are the dozen or so particles of the standard model. Individually these particles are of little direct consequence to security and defense; however, as little as a century ago the same could be argued for atoms.

The nature of matter means that atomic-scale fabrication is the finest precision possible. Atoms in general are fundamental stable building blocks. Molecular assembly is often presented as flawless replication of a product in theory. In reality, quality control would still be important. When dealing with billions of atoms there are bound to be flaws. Taking the mimicking of life further, flaws in a bottom-up construction is analogous to the growth of cancer cells. That said, any fabrication process could include an inspection capability during each step of the production process, identifying flaws shortly after they occur. Specifics of how nano-fabrication develops will determine the frequency and manner of quality control. A failure rate of 99 percent can always be justified by the right combination of low manufacturing cost and high market price for the one percent that meet standards.

Related to the nanoscale machines or manipulators assembling useful goods for home and national security are organisms that have been genetically modified (GM) to produce designer molecules. Instead of copying the chemical processes used by cells to transport and assemble atoms into molecules useful to life, this is simply reprogramming life to produce specific and tailored molecules useful to man. Much has been said about how spider silk compares favorably in strength to weight versus steel and manmade fibers such as Kevlar. Spider silk is composed of long molecule chains that combine tensile strength with flexibility. The major drawback is that spiders are not easily used in industrial processes. As of late there have been several universities and companies announcing success in genetically modifying organisms, both bacteria[17] and animals,[18] that can be easily used by industry to produce large amounts of spider silk molecule.

Nanoscale fabrication by mobile and independent atomic manipulators is simply not something for near-term consideration, a reality that even its proponents would agree with. For now, and the near future, the majority of work on nanotechnology and nanoproduction will involve top-down processes. The March 2010 assessment of the U.S. NNI recommended a focus on bringing nanotechnology to market,[19] which essentially means nanotechnology produced using developments of existing top-down processes. That said, NNI goals still include increasing funding for basic research, some of it toward better understanding on how molecules can be put together, but again no distinction is made between top-down and bottom-up processes.

3D Fabrication, Near-Term Adjunct to Nanotech or an Emerging Technology

A possible bridge toward nanoscale fabrication is the rapidly developing technology of three-dimensional (3D) printing. Three-dimensional printing, used interchangeably with the terms rapid prototyping, additive manufacturing, and desktop fabrication, is production via the laying and fixing in place of successive layers of material. Three-dimensional printers of increasingly high resolution are already operating at the micron (micrometer) scale.[20] Of course if 3D printing technology achieves layer thicknesses of less than 100 nanometers (0.1 microns) or less it, would qualify for the U.S. government's definition of nanotechnology. Between changing how industry produces complex structural items, threatening future business for package delivery companies,[21] and a growing enthusiast community constructing 3D printers for personal use, this technology is itself billed as an emerging technology with accompanying claims of being an "industrial revolution" in its own right.[22]

An important consideration with the development of these technologies is the cost-to-benefit ratio of increasing resolution, and the products that may justify the investment. Three-dimensional printing is presently being heralded for its ability to efficiently produce small production runs. For personal use this can include low cost but high-fidelity replication of household objects that one would normally only need one or two of. These items do not require nanometer precision. In industrial and military use the need for nanometer precision depends on the specific application. Some military goods, such as rifles, are produced in the thousands via existing manufacturing techniques; production via novel fabrication techniques would have to be quite low to justify their use.

One military-related product that could conceivably benefit from the development of atom-precise 3D printing and other molecular-level fabrication would be high-performance gas turbine (jet) engines. At present 3D printing is being used to produce aerospace structural components, where the ability to produce odd shapes that maximize strength while minimizing mass is being leveraged.[23] With increases to the performance of jet engines there is a tendency for more complex turbine blades, whether in material, shape, or both. Advance engine blades are formed from single crystals of metal alloy grown with great care and expense.[24] In turbine blade production, a degree of perfection is required, and today's processes are already expensive, making this an ideal starting point for economically viable molecular fabrication. It would also have security repercussions; the ability to produce military jet engines was a subject of major interest in considering Soviet military power, and today is a subject of interest when considering the military rise of China.[25]

Another possible offset to the cost of 3D printer fabrication is its potential to reduce logistical footprints. E-mail has changed how people communicate and do business, creating jobs in the IT sector, and challenging older forms of communication, such as mail and telegrams. Low marginal costs also allow for junk e-mail, spam to be a cost effective means of advertising. Among possible 3D printing futures is to use this technology to provide similar instant delivery and low marginal costs for physical products. Essentially one location would be able to e-mail a prototype or product sample to another location. As mentioned earlier this emerging capability already has raised concerns for delivery companies. Global power projection is dependent on being able to field military units far from home, and logistics for the military are more of a necessary evil than a profitable line of business. Structural spare parts, fasteners, and other simple, but necessary components, are expensive to stockpile, and/or expensive to ship as needed. There has long been military interest in using 3D printing to reduce the burden of storing these types of items. In a 1994 *Discover* magazine article, the term "fabber"[26] was used to describe this technology and noted interest from the U.S. Navy on using "fabbers" to partially replace the large parts' warehouse aboard aircraft carriers.[27] Interestingly, the article's prediction of desktop "fabbers" being available within 15 years (of 1994) seems to have been accurate.

Moving beyond specialist components and on-demand spare parts, production of nearly complete devices and equipment from 3D printers is being experimented on. Notwithstanding the need for electronics, motors, wiring, and hand assembly, 3D printers have been able to copy themselves.[28] Similarly the University of South Hampton, in the United Kingdom, flew in 2011 what is claimed to be the world's first unmanned aerial vehicle (UAV) largely produced by a 3D printer.[29]

In the United States various military organizations are known to be acquiring 3D printers, possibly to augment existing capabilities to customize and produce small batches of specialized items. Understandably, the United States Special Operations Command is not commenting on their reasons for wanting this technology.[30] Defense Advanced Research Projects Agency (DARPA) in the case of the Manufacturing Experimentation and Outreach (MENTOR) subprogram to the Adaptive Vehicle Make (AVM) program has been more open, and it is teaming 3D printers with U.S. high schools in an effort to encourage interest in science and engineering studies, and possibly directly benefit from new ideas and innovations MENTOR generates.[31]

Now it must be remembered that 3D printing does not currently have nanometer precision, and therefore cannot yet be used to produce even moderately powerful computer chips, though some mixed material printers can produce functioning circuits. Although some 3D printers can produce complex items with moving parts, anything produced larger than its "print" volume would require some assembly. Under the amusingly alarmist headline of, "Is the Navy Trying to Start the Robot Apocalypse?" online magazine *Wired* noted U.S. Navy interest in combining 3D printer and robotics technology to someday allow for autonomous assembly.[32] Object recognition by computers is still a major challenge, and the ability of artificial intelligence to comprehend and make plans in uncontrolled environments still very problematic. Dynamic assembly of loose parts is probably the more challenging problem than grafting a 3D printer onto a robot. Coordinating multiple robots toward the goal of assembling something is either again a macroscale precursor of the Drexler vision of molecular assembly, the electronic realization of "too many cooks," or something in between.

Nanotechnology and Nations

Despite being vaguely defined, nanotechnology is widely regarded as important to the equally nebulous concept of national power. The U.S. government has since 2000 been funding the NNI, which has acted as a central point of contact and collaboration between various federal agencies and others involved with many aspects of this emerging technology from fundamental research; NNI conducts research on health and safety issues involved with nanotechnology, as well as specific applications along department and agency mandates. The Department of Defense, Department of Homeland Security,

Department of Energy, NASA, and the U.S. intelligence community,[33] are all involved with the NNI.

Science, which is not of a purely military nature, has both a tendency and a need to permeate national borders. Peer review, a staple of legitimate science, requires that findings be published for others to scrutinize and ultimately replicated to prove that results were not simply chance. U.S. government agencies and companies are active with international cooperation on nanotechnology matters. This includes participation in international bodies that have studied the general implications of nanotechnology, such as State Department and Environmental Protection Agency involvement with the International Risk Governance Council,[34] as well as military organizations such as the North Atlantic Treaty Organization (NATO).[35] During the Cold War there remained some contact between scientists on both sides of the Iron Curtain even as they raced to outdo each other in the pursuit of scientific advancement. The math behind the F-117 stealth fighter can be traced directly to a scientific paper published in Russia.[36] Today, the sheer volume of scientific papers on nanotechnology being published in the People's Republic of China is being used in part to gauge this peer competitor's great interest and rapid rise in nanotechnology research.[37]

Although some of this international interest may be due to nanotechnology hype, there seems to be a shared recognition that it would be unwise to be caught on the wrong side of the "nano-divide" as far as industrial policy is concerned. Like many emerging technologies, nanotechnology may be viewed as a great leveler, and perhaps an opportunity to bypass catching up in present-day economic and military competitions, and simply move on to what is perceived to be the next competition that matters as far as national power goes. The "next competition" being a fresh playing field means that past achievements may not matter so much. As identified in U.S. Department of Defense reports, this is a major contributor to China's interest in nanotechnology research, "The PRC considers nanotechnology an area of research in which they are playing on a level field with the United States."[38] The potential of a new level playing field is the opportunity to be the ultimate winner of the next revolution in economic and military affairs.

Security and Defense Nanotechnology Opportunities

Nanotechnology is not one product or capability, but instead covers a whole host of products and capabilities. Many are simply extensions of current technology, improved through the ability to cram more capability in a given volume. On the other hand the nanoscale does open some completely new opportunities. This next section is by no means a complete list.

That the military is not alone in trying to harness this vaguely defined area of technology means that much of what will enter service will be dual-use items. An example of dual-use nanotechnology would be nanoscale structures and particles that inhibit microbial growth. These have applications that include germ-fighting hospital walls, athletic clothing that wards off odor-producing bacteria, biological weapons protection, and literally the kitchen sink, where again it is useful to prevent bacteria growth. Militaries have kitchens and hospitals where this form of passive nanotechnology may find its way into the background of military life as a disinfectant or hygiene supply. One such product, "sharkelet," mimics the nanoscale texture found in natural shark skin that prevent algae growth,[39] and had its origins in military research, and specifically U.S. Navy efforts to keep the hulls of warships and submarines clean.[40]

High-performance athletic equipment manufacturers and the aerospace industry have a common interest in high-strength lightweight materials. High-tech composites are ubiquitous elements in the design and construction of fighters and snowboards. In the near term, precisely tailored nanoscale features of fibers and other components of advance composites will allow for specific combinations of increases in tensile and compressive strength, improve elasticity, and weight reductions.

CNTs and graphene, another carbon-based nanomaterial, are two of the leading areas of interest for future high-strength lightweight materials. In concept these materials are very simple; in practice there remain many challenges to taping their properties. The bonding characteristics of carbon make it useful for the building blocks of life; organic chemistry is the chemistry of carbon compounds. Long complex molecules made up of repeated combinations of carbon and other elements make up polymers, including synthetic plastics. Carbon can bond with itself, and when such bonds are repeated among many carbon atoms, it can form buckminsterfullerene (or buckyball), a sphere of 60 carbon atoms, which is able to cage other molecules. From the buckyball came the idea of elongating the carbon-only molecular structure into the carbon nanotubes. CNT-based materials have shown remarkable tensile strength. In recent years there has been much excitement about CNT being strong enough for the space elevator concept[41]—a tether reaching from geostationary orbit to a point on the equator that may be climbed to reach space cheaply. In the near term, CNTs are being mixed into composites to reinforce their strength.[42]

Graphene is the concept of carbon bonding to carbon in a flat sheet configuration. Although only one-atom thick, this sheet would have the strength of molecular bonds keeping it intact, making it in theory many times stronger than steel.[43] That graphene is presently only available in the form of barely visible flakes means its near-term future is more likely to be found as a new material for smaller computer components due to its electrical properties[44] than as a structural material. Extreme atomic-level precision coupled with a greater understanding as to how electrons flow in such structures promises to produce superior conductors and semiconductors out of everyday carbon.

Seemingly benign and unexciting to many, material and structural sciences are behind the armor and other passive defenses that have shaped history. Naval warfare changed due to the advent of ironclad warships, neatly highlighted during the U.S. Civil War first by the near invulnerability of the Confederate ironclad CSS *Virginia* to return fire, and later in the long, arguably tactically inconclusive, slugging match between the *Virginia* and the Union ironclad USS *Monitor*. The application of iron to naval hulls both brought about obsolescence and drove development of weapons and naval tactics. The culmination of this was an arms race between armor and naval cannon, ending with the *Dreadnought* arms race between the Great Powers that preceded World War I.

Similarly personal body armor, in the form of metal suits of armor, fell out of fashion due to the weight and restrictions on movement imposed by armor becoming more of a liability. Only within the last few decades have material science produced materials such as ballistic nylon, and later Kevlar and Spectra, which have allowed for resurgence in personal body armor. None of this would have been possible if it was not for advances made over the last century in chemical and material sciences.

Beyond a technology's intrinsic value to security, there are also second-order effects that may lead to a particular capability being dysfunctional to overall security. Although current material sciences has again made viable body armor for most combat troops, critics have argued that the overt and aggressive appearance of bulky body armor have negative consequences on current operations to win hearts and minds.[45] Then there is the argument that U.S. soldiers possessing a degree of invulnerability is itself a deterrent to attack, and so on and so forth. Continued material advances that allow for less bulky and more protective body armor will change the dynamics of the arguments for and against the use of a specific type of technology. Again, these discussions would not have emerged if it was not for ongoing work on synthetic fibers. Continued advances in materials, including nanotechnology, will shape the future of these discussions.

Precisely controlled chemistry in the form of nanoscale solid reactants and nanoscale structures to contain or act as catalysts for other reactants, will unlock energetic processes from compact high explosives, to higher density batteries, and defenses against biological and chemical weapons. In reality it is not new chemistry; nanotechnology simply allows for an increase in surface area for reactions to take place. The rate, or an amount of reactions taking place at any one moment of time, is dependent on the amount of reactants in contact. Nanotechnology allows for extreme control over both the area and features of a surface allowing for the reliable and controlled reactions.

Materials that would barely smolder if exposed to fire explode at the smallest spark when in particle form. Grain silo explosions are an example of this; the correct ratio of grain dust floating in the air of a confined space, such as a partially empty silo, poses an explosion danger. Although grain and other particulate foodstuff[46] can be used as the fuel in fuel air explosives (FAEs), military research has been investigating how to get more power out of aluminum metal when in the form of nanoscale particles. Aluminum perchlorate is widely used in solid rocket motors, including those used by missiles, hobby rockets, and until the end of the U.S. Space Shuttle program, in manned spaceflight. A nanoscale aluminum particle and water ice (ALICE) propellant rocket was demonstrated in 2009 in a joint Unites States Air Force (USAF), NASA, Purdue University, and Pennsylvania State University program.[47] The 2009 report to congress on the U.S. Defense Nanotechnology Research and Development Program, listed nano-aluminum as a potential means to producing more compact warheads for bombs and missiles.[48]

The expense and interest would be wasted if the large quantities of chemical energy being unlocked via nanotechnology could not be controlled. A novel chemical reaction is worthless outside the lab if it cannot be produced in a form that only occurs on command under operational conditions. For a military product this means remaining inert when stockpiled, in transit, and preferably under all battlefield conditions barring an intentional detonation. That rapid oxidation of aluminum nanoparticles releases large amounts of energy is a reasonably expectation from all the previous use of aluminum as an energetic material; the research and development is in locking this energy in propellants and explosives that can be safely handled. The note on nano-aluminum in aforementioned 2009 report to congress on defense nanotechnology R&D was in the context of USAF research on how to stabilize the material for practical use.[49]

As the ability to control both the formation of materials and chemical reactions under industrial and operational conditions nears the molecular scale, the amount of energy that may be harnessed approaches the theoretical maximum. For explosives this means getting a bigger explosion out of a smaller charge. However, not every militarily useful chemical process involves a detonation and it is these less-spectacular reactions that may become an everyday part of military life in the near future.

A technology-dependent military, especially one with global reach, is certainly interested in compact portable sources of electricity. Many of the research being undertaken under the auspices of the NNI are energy related.

In 2006, the Department of Energy, the agency responsible for the U.S. nuclear arsenal, established the Advanced Research Projects Agency-Energy (ARPA-E), and is quite openly based on the successful model of DARPA.[50] Nanotechnology figures prominently in several ARPA-E projects, including next-generation batteries.[51] Department of Defense agencies are also separately making progress in nanoscale technologies aimed at both increasing the energy density of batteries and lowering production costs.[52] More near term are improved air-activated flameless ration heaters that use zinc[53] nanoparticles configured to react in a much slower rate than that being harnessed for explosives and rocket-propellant applications. Perhaps these applications are mundane in comparison to controversial visions of nanoscale robots infiltrating enemy forces and tearing them apart from within, but they will become part of the U.S. ability to sustain high-technology forces globally.

Chemical and quantum reactions at the small scale also constitute avenues to improve military sensors. Molecular-scale detectors are already being proposed for everything from environmental monitoring, to the diagnosis of cancer and other diseases, to the detection of chemical and biological weapon agents. All these applications have in common the detection of small number of molecules, which may bond to receptors on the sensor via chemical means. Basically these sensors are chemical traps set off by a chemical or biological process occurring on the nanoscale tripwire. Some of these reactions produce a mechanical effect, others produce optical effects, and others change the electrical properties of the material—all producing an observable effect.

Using computer chip manufacturing techniques, multiple chemical and biological detectors can be etched into what are termed lab-on-a-chip (LOC) devices. These compact laboratories are expected to facilitate continuous and ubiquitous surveillance, whether it is for medical or man-made threats. Instead of specialist hazmat teams being sent in to investigate possible nuclear, biological, and chemical (NBC) hazards, autonomous LOC sensors could be scattered across a battlefield, monitoring for these dangers. Against the threat of terrorists acquiring biological, chemical, and radiological weapons, LOC technology would allow for affordable monitoring of public spaces, ports of entry, and other soft targets for these types of attacks. LOC devices essentially replicate the ability of the sniffer dog's nose to detect traces of explosives, drugs, and other contraband.

The sensor may not actually have to be in the form of a device per se. Instead nanomaterials that react on exposure to specific molecules can be embedded into clothing and surfaces. In the presence of a hazardous material these materials would visibly change color. Moving beyond simple detection are materials that will also react to the presence of a danger. If a material can be engineered to change color on contact with a target molecule, it can also be engineered to neutralize the target molecule on contact by acting as a catalyst to its breakdown or other methods.[54] As noted earlier this is already a feature found in some commercially available clothing, though the microbes being suppressed are the less-dangerous variety that cause body odor.

Electronic devices in general would be a given for nanotechnology, especially as the cutting edge is already at the nanoscale. That said, the nanoscale actually represents a barrier to the continued shrinking of electronics. Phenomena that only manifest at these small scales are producing challenges that were not present only a few years ago. At the nanoscale, quantum effects have a bearing on how a device or system operates, disrupting the orderly world of electrical engineering. Quantum tunneling, where particles cross what would under other physics be impenetrable barriers, becomes a concern once data-bearing circuits are miniaturized beyond a certain size. Electrons that would constitute signals in separate data pathways would leap from one another in an uncontrolled manner through quantum effects, corrupting the data that was being transmitted. On the other hand, quantum effects may be harnessed as in the case of ongoing work to produce a viable quantum computer. Quantum computers are regularly promoted as being able to, in theory, break any digital encryption scheme, and therefore are generating vast amounts of real-world interest from intelligence agencies and other national security entities.

Besides manipulating digital information, devices crafted via the same processes used to make computer chips will be able to interact with the physical world in more ways. Display and optics technology are among the easier means for small devices to interact with the world. While the media have been recently captivated by the prospect of "invisibility cloaks" made from metamaterials—materials with nanoscale features engineered to bend and shape EM radiation so that it flows around an object passively—the problem with metamaterials is that thus far they only work on specific wavelengths of light and can only enclose small volumes. The ability to selectively steer parts of the EM spectrum, however, has applications in optics, computing, and other devices. Near-term optical camouflage is more likely to be produced by embedding in the exteriors of vehicles and in uniforms very small electronic elements that change color, and emitting the correct amount of light so as to blend in with ambient conditions. The smaller the active elements are, the more effective this form of camouflage is—a skin of active nanoelectronic optical elements representing the ultimate form of this capability. Active camouflage has run into problems with real-world engineering, costs, and acceptance (counter-illumination was tested during World War II[55]), but is slowly making it to market in lower-resolution forms, and across limited portions of the EM spectrum.[56] It is possible that the cost and robustness challenges will be overcome in related active display technology that is used in billboards and large-scale outdoor displays, meaning active camouflage may come from a dual-use product.

The current (or last, depending on the pundit) Revolution in Military Affairs is information hungry, leading to interest in producing a smart dust, small electronic sensor platforms networked together by wireless, to cover the battlefield. The smart dust concept is attributed to Dr. Kris Pister,[57] and though it is now being marketed as general networking technology, it had its roots in a DARPA project.[58] The smart dust concept is somewhat related to ubiquitous computing, another potential emerging technology, where every-day objects would be equipped with computer processing and be networked to be capable of forming a "network of things." Blanketing the battlefield with sensors and computers will go a long way toward pushing back the fog of war, but may also increase friction due to information overload.

Programmable matter is for some the ultimate computing input and output device, or more correctly collection of devices. Basically programmable matter, as worked on by DARPA,[59] is a collection of particles that each have the ability to act as structural material, electronic/optical device, locomotion, and controlled attachment. As pixels form images on television and computer screens, collections of modular nanoscale robots would link together to form 3D objects. Another term used for this concept is claytronics.[60] This is in all likelihood a long-term vision; aside from the nanoscale challenges of constructing one of these multipurpose robots, let alone the millions needed, would be the computational task of coordinating them.

In between the immobile smart dust and swarms of modular robots of programmable matter are projects aiming to produce bird- or insect-sized robots for surveillance and military applications.[61] The small robot form of drone strikes would, however, be somewhat challenging from a policy standpoint— at the size of an insect or below, not much of an explosive warhead could be delivered. Lethal payloads that could be delivered by an insect-sized weapons platform seem to be problematic with the 1993 Chemical Weapons Convention (CWC), and the 1972 Biological Weapons Convention (BWC), or both.

Nanoscience and Engineering: As the Latest World-Shattering Worry

Although there is always some opposition to new technologies, let alone technologies with military applicability, nanotechnology seems to have gained extra attention. Much of this stems from recognition that the emerging technology of nanotechnology is likely to change the world. The economic impact is clear; at the very least nanotechnology will bring many novel products to market, and if the more radical visions of nanotechnology bear fruit then the very foundations of manufacturing will undergo a revolution. From an international-relations perspective, massive changes to economic and systemic norms are generally something to be concerned with as these are contributors to international tensions. With any change to the status quo there are always winners and losers, and therefore if one accepts a particular technological change, then it would be useful to consider how to come out ahead.

Nanotechnology is often compared to the nuclear sciences—a powerful technology, once hyped as the future, but now facing scrutiny over its health effects, as well as its potential for destruction. In the 1950s, civilian nuclear technology seemed to be promoted for all areas of civilian life, including powering automobiles. The nuclear arms race and worries over the safety of nuclear power led to a backlash against this technology. This backlash against nuclear technology in general has in recent years been mitigated by the perceived need for energy sources that do not generate pollution in the form of greenhouse gases.

Both the pro- and anti-nanotechnology sides of the controversy can also find parallels in the scrutiny and arguably unwarranted hysteria, with that being faced by GM foods in Europe—rightly or wrongly, in a scientific or public policy sense, GM foods have gained an emotive stigma that has thus far limited what was once a promising area of research. Opponents of nanotechnology and of GM foods often place emphasis on the risks these technologies present, including the threat of presently unknown risks.[62] Many of the same groups opposed to GM foods, "see nanotechnology as the next natural target."[63] Proponents of nanotechnology on the other hand are keen to avoid having their field succumbing to the same hysteria, backlash to overselling, and marketing missteps of the GM food industry.[64] Nanotechnology, like GM agriculture, and nuclear technology for that matter, has risks, benefits, and a great potential for emotion to cloud the debate.

Appropriate risk assessment is now part of the debate that will shape how nanotechnology enters society. With many of the desired properties of nanotechnology still in the realm of theory and speculation, the undesirable qualities are subject to similar unknowns. It is quite possible that many harmful, if not deadly, effects will only become apparent over time. Unlike the early days of nuclear science, where radiation was hyped as a cure all, nanotechnological promises are accompanied with scrutiny and a degree of caution.

There are two schools of thought on nanomaterials safety: nanomaterials are often forms of commonly used elements and compounds, with known health and safety properties; or they represent totally new materials that must face consideration as such. One of the cited promises of nanotechnology is "unique phenomena" that only become manifest with the small sizes of the particles and structures involved. At this scale, it is feared that unexpected and potentially unhealthy phenomena would occur. Some have even argued that this potential for harm is too great and call for a slowing of all nanotechnology development until all such effects can be studied.

In the rush to find applications for new materials, there have been several examples of seemingly useful materials proving hazardous to human health over time. A perhaps more relevant comparison of the potential short-term hazards of nanotechnology is the natural fiber asbestos and other potentially dangerous industrial substances.[65] All small particles have the potential to enter the body and cause health effects ranging from simple irritation to increased risk of cancer. By virtue of being at the scale of cellular organelles, or smaller, there is concern that nanoscale particles and fibers would have all manners of unexpected effects on human health. Asbestos is among the most notorious, being the subject of numerous lawsuits over its carcinogenic effects.

The very same properties that may allow risky materials to penetrate living cells are also being investigated for their potential as mechanisms to deliver treatments for cancers and other illness. Therefore it is largely a matter of context and control, or lack of control, that is the problem. It must be remembered that it is not nanomaterials in general that are hazardous, but that under specific circumstances materials, such as CNTs[66] and gold-based quantum dots,[67] may have unwanted health effects. A consequence of this very real concern over unwanted, and especially unexpected, reactions between nanomaterials and living tissue is investment in research to seek out these effects and mitigate them when possible.[68] All industries, from heavy industry, to food production, and even call centers, have occupational hazards.

In the military realm, depleted uranium (DU) is the contemporary material under the spotlight due to its possible health effects. DU is slightly radioactive, though only dangerous internally. DU is also a heavy metal, though its effects are less than that of common industrial metals, cadmium, lead, and mercury.[69] DU is a very hard material, making it very useful as both armor plate and as kinetic-energy penetrators for defeating armor. Unlike the ordinary U.S. workplace, the battlefield exposes DU and other novel materials to the extremes of combat. The use of DU in warfare may result in the formation and release of possibly dangerous compounds and particles. DU in addition to being extremely hard also burns very hot. Although the pyrophoric (self-igniting) properties of DU are militarily useful secondary effects,[70] the toxic by-products have come under criticism from both the international community[71] as well as U.S. personnel who may be exposed to its effects in the course of their duties.[72] It must, however, be noted that there is at present still no conclusive link between battlefield use of DU and health issues afterward.[73]

Like DU, the great promises of nanotechnology outweigh the many known and speculated on risks. Being overcautious is a risk in itself as nations are vying to create favorable conditions to become the leader in what is widely thought to be a critical technology for the foreseeable future. Although it is fair to say that risk assessment for military applications is different from that for civilian use, most would consider it callous, possibly treasonous, for policy makers in most Western nations to carelessly poison their own service members. That this has arguably happened before with the legacy of the defoliant Agent Orange is stark warning for many. It must, however, also be remembered that some security opportunities are so great that, even if hazardous, they must be pursued.

As with many military emerging technologies there are those that wish to bring into place treaties to limit its potential for damage. This extends up to the point of proscribing some areas of research, including the entire "radical vision" of nanotechnology. In another parallel with nuclear technology, nanotechnology in more extreme predictions, is not only a weapon of mass destruction (WMD), but is argued as having greater potential than nuclear arms to bring about the end of civilization. This line of thought has been put forth by Ray Kurzweil, who generally views nanotechnology as beneficial.[74] Many areas of nanotechnology remain purely in the theoretical realm today, and opponents of weaponized nanotechnology and even nanotechnology in general, see a need to halt progress well before the worse-case consequences are within reach.[75]

Among the worse-case consequences is the gray goo scenario. A viable universal molecular assembler would also have to be capable of disassembling matter as an initial step in reconstituting it into a desired product; if this capability is applied in a coercive manner, the destruction of a nation's military equipment, infrastructure and/or personnel, for instance, this industrial tool becomes a weapon. The universal molecular assembler, if viable, is dual use. Related to the intentional use of the theoretical molecular assembler as a weapon, is the possibility of a runaway "gray goo" scenario due to malfunction.[76] A self-replicating assembler may have from the very beginning flawed software or flawed design leading to loss of human control. These flaws may even arise due to errors in the self-replication process, when a process is replicated millions of times, there must be some percentage of errors.

Another position is that this emerging technology should guided by an all-inclusive international body for the purposes of directing it toward what "civil society" would call global benefits.[77] Although some may find this idealism inspiring, it must be remembered that compared to debates over the health effects of nanotechnology, let alone the threat of nanotechnological "gray goo," there has been several magnitudes more controversy, and bloodshed, over what constitutes worthwhile global benefits and the proper order of things in the world. This includes the reality that an absolute end to global disparity may not necessarily be worth the price. Moreover there it would be irrational for any powerful nation state to voluntarily hamstring itself in security or economic terms purely out of charity or sentiment.

Staying firmly grounded in the real world of international relations, where knowledge of another international actor's motives and plans must always be treated as imperfect, nations have in a few narrow instances found utility in the banning of certain military technologies—though not without cases of cheating. Biological weapons, another existing WMD the world has to deal with, are not only subject to particular abhorrence by the world at large, but are also specifically banned by the 1972 BWC.[78] Now the caveat is that verification methods for the BWC were not part of the original 1972 treaty; indeed treaty verification is often a sticking point especially when dealing with governments with poor reputations for transparency. Admissions by the Russian government[79] confirm the existence of a biological weapons program well after the Soviet Union (the signatory state to which Russia is the successor) was subject to the BWC.

The nanotech assembler once created would be hard to monitor. As a dual-use technology it would be impossible to prevent it from being reprogrammed for military use. With self-assembly, only a small sample, potentially only one weaponized assembler, would need to be kept hidden for the treaty to be rendered meaningless. One assembler could through exponential growth quickly rebuild swarms of weaponized assemblers. The ease at which one could cheat with nanotechnology would seem to make any treaty pointless.

This leads to the broader debate over whether there can there be instances where an arms race is preferable to an imperfect treaty? Common to both existing biological weapon treaties and suggested nanotechnology limits are fears that such regimes may actually decrease security by preventing nations from investigating effective countermeasures. To understand the threat, both from intentional and accidental destructive release of self-replicating nano-assemblers into an environment, requires a stock of such devices, but under controlled circumstances. This is analogous to the remaining samples of the small pox virus, kept in the United States and Russia for research purposes. Small pox was once a major killer, and even used as a biological-warfare agent; through international efforts it was the first human disease to be totally eradicated, outside of samples kept controversially in the United States and Russia. Among the reasons for keeping these samples is continued research into countermeasures against it as a potential bioweapon if it were ever to be reintroduced to the world.

Then there are treaties that have become obsolete, or dysfunctional, for technological or political reasons. The Bush administration's withdrawal from the 1972 antiballistic missile (ABM) treaty in 2001 was made on the basis that this treaty would hinder U.S. security by preventing it from deploying limited missile defense systems to counter the threat from rogue states possessing rudimentary intercontinental ballistic missiles (ICBMs) and nuclear weapons technology. During the ABM treaty's inception, it was argued that missile defense efforts by the United States or the Soviet Union, targeted against the other's strategic missiles, would be a destabilizing factor. In this sense the role of the ABM treaty, as originally envisioned, was to protect the effectiveness of strategic nuclear missiles. Nuclear deterrence arguably kept the Cold War stable, where the actual limitation, let alone banning, of offensive weapons between the United States and Soviet Union (succeeded by Russia) has been difficult to achieve. The Cold War is over, but new nuclear threats are potentially only years away. Moreover a limited missile defense is not capable of changing the balance of terror, that is, nuclear deterrence between great powers.

In the hype surrounding the "gray goo" scenario, it has been suggested that the ultimate countermeasure to a cloud of nanoscale assemblers programmed to attack, or simply run amuck, would be another cloud of nanoscale assemblers programmed to attack the threat cloud. Under this argument to have options in defending against a rogue swarm of assemblers, a nation cannot be prevented from constructing its own swarm of assemblers.

Adding to the many challenges of getting comprehensive agreement between nations to limit nanotechnology's more destructive aspects is the fact that the promise of nanotechnology goes beyond the empowerment of largely responsible great powers. Although nations may be able to agree, or at least be deterred from using nanotech weapons, individuals may not. After all it may take just one assembler programmed to bring into being a destructive assembler swarm, if it was possible, to be unleashed on the world. A handful of fanatics with money for flying lessons were able to bring terror to the U.S. homeland. A single maladjusted individual with access to a nanotech assembler would be able to threaten the world. Utopian claims of the radical "vision" of nanotechnology being democratizing, putting the engines of a new industrial revolution in the hands of people everywhere, would seem to be somewhat naïve in the face of the capacity for evil found in some individuals.

Although it must be stressed that the nanotech assembler is still a far-off prospect, possibly not even commercially viable (rendering the gray goo fears similarly remote),[80] other forms of manufacturing related to nanotechnology bring along their own worries. In one critical respect, proliferation, the comparison with nuclear technology breaks down. Nuclear technology in general is difficult to master and, due to its obvious military applications, highly regulated. As a large-scale industrial process, the refinement of nuclear materials into fuel and weapons' grade materials is difficult, though not impossible, to hide. Since the mid-1940s only a small number of nations have been able to acquire the full spectrum of nuclear capabilities. International regimes exist, providing both incentives and penalties, to stop additional nations from acquiring nuclear weapons.

Three-dimensional fabrication hints at the possibility of personal factories, or fabricators, able to produce anything for which the necessary materials are available. Aside from disrupting entire sectors of the global manufacturing economy, these small factories if programmed to produce military goods would allow weapons' proliferation on a scale unimaginable today. Not only small arms, but also sophisticated weapons such as antiaircraft missiles would be within grasp of many subnational threats such as terrorists and criminal elements. All the existing regimes limiting access to many advance weapons could be circumvented once one of these fabricators made its way into a rogue state or organization. Once in place, all it would need would be raw materials and the exquisitely detailed blueprints that this technology requires to produce fully working copies of any weapon within its manufacturing capabilities.

A nation's ability to produce effective armaments is part of its strategic capabilities. This ability is not purely manufacturing, or purely defense research and development. Instead it is achieving the correct balance between cutting-edge military research and the ability to produce effective quantities of usable weapons. This balance, the seemingly eternal quantity-versus-quality debate, changes throughout time. The World War II lasted for years and throughout the conflict both sides were presented with difficult choices over continuing with tried and trusted weaponry or gambling with new technology. The United States' manufacturing capability became, as President Roosevelt put it, the "arsenal of democracy"; however, the United States also hosted the Manhattan Project to develop a bomb based on what was for its time cutting-edge physics. The Cold War that followed was punctuated with a need for constant vigilance; a major systemic war between the West and the communist spheres of influence was expected to be a brief and sharp affair leaving little time to build up. Additionally, weapons were becoming more complex—shifting more emphasis toward a nation's ability to produce high-technology items. Many arguments are made about the "defense death spiral"—where the cost and complexity of newer platforms result in fewer of these platforms entering service and, consequently, being of less utility over time.

Wide-scale proliferation of small-scale fabrication would potentially allow for both mass and quality to be available at low cost. Combined with espionage, even the most secret of high-tech weapons would be available for the price of the stolen data. Large investments in military research and development would be meaningless if the data could not be secured. This was one of the potential pitfalls of nanotechnology discussed by Drexler in his 1993 book, *Unbounding the Future: The Nanotechnology Revolution*, with the example of hypothetical Singapore, in the real world a small city state, though one with a modern high-tech economy, using nanotechnology to emerge suddenly from out of nowhere as a military superpower to challenge the United States and Japan.[81]

Although the use of Singapore was used to highlight nanotechnology's capacity to rapidly shift conventional power balances, the scenario presented by Drexler hinted also that warfare would become borderless. If personal fabrication technology became ubiquitous, and capable of producing hi-tech goods, then small teams, whether they be spies or terrorists or simply disaffected citizens who had been swayed by a hostile ideology, and armed only with the plans for a weapon such as a UAV or even the triggering mechanism for an improvised explosive device (IED), could attack a country with open borders, such as the United States, from within. With terrorism using modern social networking tools to spread not only their ideology, but also the technical knowledge to carry out an attack[82] advance fabrication would only be tool for rapidly assembling an attack seemingly from out of nowhere.

Nanotechnology, for all the worry about it falling into the wrong hands and debate over how it should be used, is still ultimately a tool—potentially powerful in the extreme but still a tool nonetheless. Molecular-level manipulation touches on all areas of industry, warfare, and life, meaning it will certainly be in demand, and hard to suppress. It is only an enabler for human motivations; so like all technologies once it is available, it will be perhaps better to worry less about the tools, and more about the motivations that drive its use.

Notes

1. *The National Nanotechnology Initiative, Supplement to the President's FY 2012 Budget,* http://www.nano.gov/sites/default/files/pub_resource/nni_2012_budget_sup plement.pdf.

2. Richard A. L. Jones, *Soft Machines: Nanotechnology and Life* (New York: Oxford University Press), 55.

3. Kevin Bullis, "Climbing Walls with Carbon Nanotubes," *Technology Review,* June 25, 2007, http://www.technologyreview.com/computing/18966/?mod=related.

4. The van der Waals force is in turn simply the net magnetic forces between atoms. Atoms have north and south poles; when two solid materials are brought close enough together, surface atoms are in close enough proximity that magnetic attractions between the atoms are appreciable.

5. Kevin Bullis, "Climbing Walls with Carbon Nanotubes," *Technology Review,* June 25, 2007, http://www.technologyreview.com/computing/18966/?mod=related.

6. Defense Advanced Research Product Agency, "Z-Man," http://www.darpa.mil/ Our_Work/DSO/Programs/Z_Man.aspx.

7. Foresight Institute, "About Nanotechnology," http://www.foresight.org/nano/index.html.

8. Richard A.L. Jones, *Soft Machines: Nanotechnology and Life* (New York: Oxford University Press, 2004), 83–85.

9. Ibid., 3.

10. Dónal P. O'Mathúna, *Nanoethics: Big Ethical Issues with Small Technology* (United Kingdom: Continuum, 2009), ix.

11. Ibid., 22.

12. Jon Evans, "Manufacturing the Carbon Nanotube Market," *Chemistry World*, November 2007, Vol. 4, No. 11, http://www.rsc.org/chemistryworld/Issues/2007/November/ManufacturingCarbonNanotubeMarket.asp.

13. IBM, "IBM Spelled with 35 Xenon Atoms," September 28, 2009, http://www-03.ibm.com/press/us/en/photo/28500.wss.

14. Arthur C. Clarke, *2010: Odyssey Two* (New York: Del Rey, 1982), 334.

15. K. Eric Drexler, *Unbounding the Future: The Nanotechnology Revolution* (Foresight Institute, 1983), http://www.foresight.org/UTF/download/unbound.pdf.

16. Douglas Mulhal, *Our Molecular Future—How Nanotechnology, Robotics, Genetics, and Artificial Intelligence Will Transform Our World* (Amherst: Prometheus Books, 2002), 277–81.

17. European Science Foundation, "Big Molecules Join Together Will Lead to Better Drugs, Workshop Found," February 20, 2008, http://www.esf.org/media-centre/ext-single-news/article/new-understanding-of-how-big-molecules-join-together-will-lead-to-better-drugs-synthetic-organic-ma.html.

18. Rebecca Boyle, "How Modified Worms and Goats Can Mass-Produce Nature's Toughest Fiber," *Popular Science*, October 6, 2010, http://www.popsci.com/science/article/2010–10/fabrics-spider-silk-get-closer-reality.

19. *Report to the President and Congress on the Third Assessment of the National Nanotechnology Initiative*, March 12, 2010, http://www.whitehouse.gov/sites/default/files/microsites/ostp/pcast-nano-report.pdf.

20. Filton, "The Printed World," *The Economist*, February 10, 2011, http://www.economist.com/node/18114221?story_id=18114221.

21. Ibid.

22. Duncan Graham-Rowe, "'Gadget Printer' Promises Industrial Revolution," *New Scientist*, January 8, 2003, http://www.newscientist.com/article/dn3238.

23. Filton, "The Printed World," *The Economist*, February 10, 2011, http://www.economist.com/node/18114221.

24. Lee S. Langston, "Crown Jewels: These Crystals Are the Gems of Turbine Efficiency," *Mechanical Engineering Magazine*, February 2006, http://memagazine.asme.org/Articles/2006/February/Crown_Jewels.cfm.

25. Bradley Perrett, "China Aims For Gas Turbine Catch-Up," *Aviation Week*, September 30, 2011, http://www.aviationweek.com/aw/generic/story_generic.jsp?channel=awst&id=news/awst/2011/10/03/AW_10_03_2011_p22–376290.xml&headline=China%20Aims%20For%20Gas%20Turbine%20Catch-up.

26. Scott Faber, "Printing in 3-D," *Discovery*, September 1994, http://discovermagazine.com/1994/sep/printingin3d426/?searchterm=3d.

27. Ibid.

28. Ben Rooney, "The 3D Printer That Prints Itself," *Wall Street Journal*, June 10, 2011, http://blogs.wsj.com/tech-europe/2011/06/10/the-3d-printer-that-prints-itself/.

29. Paul Marks, "3D printing: The World's First Printed Plane," *New Scientist*, August 1, 2011, http://www.newscientist.com/article/dn20737-3d-printing-the-worlds-first-printed-plane.html?full=true.

30. Adam Rawnsley, "MakerBot Commandos: Special Ops Seek 3D Printer," *Wired*, August 12, 2011, http://www.wired.com/dangerroom/2011/08/special-ops-meets-makerbot-commandos-want-3d-printer/.

31. DARPA, "Adaptive Vehicle Make (AVM)," http://www.darpa.mil/AVM.aspx.

32. Adam Rawnsley, "Is the Navy Trying to Start the Robot Apocalypse?" *Wired*, March 3, 2011, http://www.wired.com/dangerroom/2011/03/navy-robot-apocalypse/.

33. NSET's Participating Federal Partners, http://nano.gov/partners.

34. International Risk Governance Council, *Nanotechnology Risk Governance*, 2007, http://www.irgc.org/IMG/pdf/PB_nanoFINAL2_2_.pdf.

35. North Atlantic Treaty Organization, *179 STCME 05 E—The Security Implications of Nanotechnology*, http://www.nato-pa.int/default.asp?SHORTCUT=677.

36. Ben Rich and Leo Janos, *Skunk Works* (New York: Little, Brown and Company, 1994), 19.

37. Department of Defense; Director, Defense Research and Engineering, *Defense Nanotechnology Research and Development Program: Report to Congress*, December 1, 2009, http://nano.gov/node/621.

38. Ibid.

39. Sharklet Technologies, Inc. "Technology," http://www.sharklet.com/technology/.

40. Ibid.

41. NASA, "Audacious & Outrageous: Space Elevators," September 7, 2000, http://science.nasa.gov/science-news/science-at-nasa/2000/ast07sep_1/.

42. J. Makar, J. Margeson, and J. Luh "Carbon Nanotube/Cement Composites—Early Results and Potential Applications," http://www.nrc-cnrc.gc.ca/obj/irc/doc/pubs/nrcc47643/nrcc47643.pdf.

43. Alex Hudson, "Is Graphene a Miracle Material?" *BBC*, May 21, 2011, http://news.bbc.co.uk/2/hi/programmes/click_online/9491789.stm.

44. Alexis Madrigal, "Scientists Build World's Smallest Transistor, Gordon Moore Sighs with Relief," *Wired*, April 17, 2008, http://www.wired.com/wiredscience/2008/04/scientists-buil/.

45. Tim Blackmore, *War X: Human Extensions in Battlespace* (Toronto: University of Toronto Press, 2005).

46. Popular science entertainment program Mythbusters demonstrated the effectiveness of nondairy coffee creamer as the fuel in a fuel air reaction.

47. "AFOSR and NASA Launch First-Ever Test Rocket Fueled by Environmentally-Friendly, Safe Aluminum-Ice Propellant," August 21, 2009, http://www.wpafb.af.mil/news/story.asp?id=123164277.

48. Department of Defense; Director, Defense Research and Engineering, *Defense Nanotechnology Research and Development Program: Report to Congress,* December 1, 2009, http://nano.gov/node/621.

49. Ibid.

50. Department of Energy, Advanced Research Projects Agency—Energy, "About," http://arpa-e.energy.gov/About/About.aspx.

51. *The National Nanotechnology Initiative, Supplement to the President's 2012 Budget,* http://www.nano.gov/node/602.

52. Ibid.

53. Department of Defense; Director, Defense Research and Engineering, *Defense Nanotechnology Research and Development Program: Report to Congress,* December 1, 2009, http://nano.gov/node/621.

54. Ibid.

55. David Hambling, "Cloak of Light Makes Drone Invisible," *Wired,* May 9, 2008, http://www.wired.com/dangerroom/2008/05/invisible-drone.

56. "Tanks Test Infrared Invisibility Cloak," *BBC,* September 5, 2011, http://www.bbc.com/news/technology-14788009.

57. Dust Networks, "Company Background," http://www.dustnetworks.com/about/company_overview.

58. "What Is Smart Dust, Anyway?" *Wired,* June 2003, http://www.wired.com/wired/archive/11.06/start.html?pg=10.

59. Henry S. Kenyon, "Programmable Matter Research Solidifies," *Signal,* June 2009, http://www.afcea.org/signal/articles/templates/Signal_Article_Template.asp?articleid=1964.

60. Seth Copen Goldstein, Jason D. Campbell, and Todd C. Mowry, "Programmable Matter," *IEEE Computer* 38(6): 99–101, June 2005, http://www.cs.cmu.edu/~claytronics/papers/goldstein-computer05.pdf.

61. P. W. Singer, *Wired for War* (New York: The Penguin Press, 2009), 118.

62. Toby Shelly, *Nanotechnology: New Promises, New Dangers* (Nova Scotia: Fernwood Publishing), 82–86.

63. Richard A. L. Jones, *Soft Machines: Nanotechnology and Life* (New York: Oxford University Press), 5.

64. Ibid.

65. Dónal P. O'Mathúna, *Nanoethics: Big Ethical Issues with Small Technology* (United Kingdom: Continuum, 2009), 69.

66. National Institute for Occupational Safety and Health, Centers for Disease Control and Prevention, "Nanotechnology," http://www.cdc.gov/niosh/review/peer/HISA/nano-pr.html.

67. National Institute for Occupational Safety and Health, Centers for Disease Control and Prevention, "Approaches to Safe Nanotechnology," 2009, http://www.cdc.gov/niosh/docs/2009–125/.

68. National Science and Technology Council Committee on Technology, Subcommittee on Nanoscale Science, Engineering, and Technology, "National Nanotechnology Initiative Environmental, Health, and Safety Research Strategy," October 2011, http://www.nano.gov/sites/default/files/pub_resource/nni_2011_ehs_research_strategy.pdf.

69. D. R. Goodman "Nephrotoxicity. Toxic Effects in the Kidneys," in *Industrial Toxicology Safety and Health Applications in the Workplace*, ed. P. L. Williams and J. L. Burson (New York, NY: Van Nostrand Reinhold Company, 1985), 106–22. Quoted in CDC http://www.atsdr.cdc.gov/ToxProfiles/tp150-c3.pdf.

70. Under the UN Convention on Conventional Weapons, Protocol III, a secondary incendiary effect would not have DU ammunition classified as an incendiary weapon. The primary effect of DU ammunition is armor penetration.

71. Avril McDonald, "Depleted Uranium Weapons: The Next Target for Disarmament?" *Disarmament Forum*, Vol. three, 2008, http://www.unidir.org/pdf/articles/pdf-art2757.pdf.

72. Naomi Harley, et al. "A Review of the Scientific Literature as It Pertains to Gulf War Illnesses Volume 7: Depleted Uranium" (RAND Corporation, 1999), http://www.rand.org/pubs/monograph_reports/MR1018z7.html.

73. Ibid.

74. Ray Kurzweil, *The Age of Spiritual Machines* (New York: Viking, 1999), 141–42.

75. Bill Joy, "Why the Future Doesn't Need Us," *Wired*, April 2000, Issue 8.4, http://www.wired.com/wired/archive/8.04/joy.html.

76. Sean Howard, "Nanotechnology and Mass Destruction: The Need for an Inner Space Treaty," *Disarmament Diplomacy*, No. 65, July–August 2002, http://www.acronym.org.uk/dd/dd65/65op1.htm?.

77. Toby Shelley, *Nanotechnology: New Promises, New Dangers* (Nova Scotia: Fernwood Publishing, 2006), 131–50.

78. Gregory D. Koblentz, *Living Weapons* (Ithaca: Cornell University Press, 2009), 7.

79. Brigadier General Russ Zajtchuk, editor, et al., *Medical Aspects of Chemical and Biological Warfare* (Bethesda: Office of The Surgeon General, Department of the Army, United States of America, 1997), http://www.bordeninstitute.army.mil/published_volumes/chemBio/chembio.html.

80. Paul Rincon, "Nanotech Guru Turns Back on 'Goo'," *BBC*, June 9, 2004, http://newsvote.bbc.co.uk/mpapps/pagetools/print/news.bbc.co.uk/2/hi/science/nature/3788673.stm.

81. K. Eric Drexler. *Unbounding the Future: The Nanotechnology Revolution* (Foresight Institute, 1983), http://www.foresight.org/UTF/download/unbound.pdf.

82. Michael Holden, "Make Cupcakes, Not Bombs," *Reuters*, June 3, 2011, http://uk.reuters.com/article/2011/06/03/uk-britain-mi6-hackers-idUKLNE75203220110603.

Biotechnology

The field of biotechnology presents new opportunities and dangers in the realm of national security. Biotechnology is essentially technology meant to augment our control over life, and possibly someday even death. The application of the life sciences is not a new thing—agriculture and medicine have several millennium of history. What sets apart the biotechnology being speculated on today is its potential for rapid and precise control over a growing number of biological processes. The promise of biotechnology has the potential to reshape the lives of individuals, their descendants, and civilization wholesale.

Military biotechnology reinforces the primacy of people at the core of military affairs. It is for human needs that wars and politics by less-drastic means are conducted. With strong artificial intelligence either in the far future or simply an impossibility, human intelligence will still be calling the shots in the near term and probably the medium term as well. Warfare is not an antiseptic video game, and the consequences of war are suffered by people—the traumas suffered by the warfighter, those caught in between, as well as those back at home. Even while other emerging technologies may be increasing the distances at which the enemy may be engaged, the effective use of military force cannot overlook the human warfighter.

Biotechnology, even without military applications, is a controversial technological area. For many, biology and medicine are unsettling fields due to the visceral nature of the subject. The life sciences in general involve wet and seemingly unpredictable things. Many see advance biotech as simply an exercise in hubris, with technological control over life at the degree promised unobtainable. Others question whether humanity can be trusted with such power over life itself—something that historically has been regarded as the power of deities.

Thanks to many past and ongoing high-profile cases of unintended medical and agricultural side effects, there are those who doubt that there will be true mastery over biology. Nature has repeatedly thrown up unexpected challenges when a medical breakthrough seemed within reach. Possible unintended side effects of militarized biotechnology range from simple personal injury, all the way up to extinction-causing epidemics. Some forms of genetic treatment and modification affect reproductive cells, meaning that these changes would be passed on to any children conceived after the genetic adjustments took hold. With society becoming increasingly risk adverse there are some areas of biotech that society may decide are simply not worth pursuing.

There are of course elements of civil society that would automatically label militarized use of any technology unwise. Some find specific cause for concern in biotechnology's potential to infringe on individual free will and human dignity. Past experiences with nations attempting to improve humanity have led to all manner of nightmarish policies from institutionalized discrimination, to wholesale murder. Noted defense and international relations scholar Martin van Creveld wryly comment that many of the technologies discussed in this chapter would be applicable to the elimination of war—via the end of individual freedom:

> A completely automated thought-reading machine—for nothing less would do—would have to be hooked up with the human brain and capable of influencing it by chemical or electrical means. Robots would have to control men, men themselves turned into robots. We find ourselves caught in a cross between Huxley's *Brave New World* and Orwell's *1984*. So monstrous is the vision as to make even war look like a blessing.[1]

Some in the field of medical ethics are troubled about the use of medical knowledge to improve the effectiveness of weapons. Although there is probably not much angst found in the research and development needed to produce a logistically more convenient field ration, there is great debate over using an increasing understanding of human biology to make a more effective kill mechanism. Somewhere in between is the application of life-sciences knowledge to less-than-lethal weapons. At its core are philosophical questions over what constitutes harm, and how such high ideals mix in with real-world practicalities of national security.

Another sign of tension between scientific ethics and practical security was the recent debate over U.S. government's attempts to prevent open publication of research on the H5N1 virus, specifically how it can be intentionally modified to be more easily transmittable.[2] Legitimate science requires free and open publication; however, with advance biotechnology tools there are fears that this knowledge may be used for bioterrorism.[3]

Arguably in this case, it seems the needs of national security led to a more cautious stance and the high ideals of science are perhaps the more reckless position.

The convergence of biotechnology with other emerging technologies such as nanotechnology and advance computing is leading to a future that will be hard to predict. Although some futurists are happy to paint visions of a future of longer life and better health, others are not so sure. Old threats such as biological weapons will have in the emerging technologies new avenues for development and deployment. As these very same technologies are providing countermeasures, it is unclear if biotech and other emerging technologies will lead to a more or less secure world. By definition, the technological singularity that some have predicted as the ultimate result from the convergence of today's emerging technologies is a point where the future will become impossible to predict due to an accelerated pace of technological development and scientific discovery. That a technological singularity is not the only possible outcome from the emerging technologies only adds to the uncertainly.

That there will be convergence of these areas is the only certainty. Nanotechnology exits at a scale that overlaps with biological constructions such as components of living cells. This means that nanotechnology is a possible means to directly adjust the function of living creatures down to the level of cells. Rudimental linkages between electronics and the human nervous system have already been implanted. Even without direct implantation, there are continuous developments in the area of noninvasive human-machine interfaces borne out of research into how the people can make use of hard technology—and how such technology changes how people work, play, and even fight. Biotechnology on the other hand is being harnessed to produce nanoscale products and may in future allow for computers to continue shrinking beyond barriers to conventional integrated circuit fabrication. Even the more specifically military- and aerospace-related emerging technologies have their links to biotech. Directed energy weapons converge with biology in the form of understanding how various forms of directed energy may disrupt parts of the body, such as the nervous system, to produce less-than-lethal effects. In the event that space access were to become a ubiquitous military capability, organizations such as the United States Marine Corp have already started to look again at the requirements for prompt global troop deployment via suborbital spaceflight[4]—where among the challenges would be life support. Technology-enhanced soldiers, although not unique to the space deployment concept,[5] would be a critical requirement simply due to the logistical costs involved. Overall it is perhaps the training and support improvements borne out by biotechnology that will provide the most military utility.

Biotech: The Familiar, the Mundane, the Prelude to Unsettling Things?

Biology is fundamentally chemistry, and chemistry at its most basic level is physics. The function of every molecule critical to life, deoxyribonucleic acid (DNA), proteins, hormones, neurotransmitters, and the adenosine triphosphate that powers cells all can be understood in terms of hard physics. Now living cells are complex systems, but with the convergence of biotechnology with computer technology much of the computational burden of unlocking life's secrets has become manageable. The completion of the Human Genome Project in only 13 years[6] was only possible with the exponential growth of computer technology.[7] As mentioned in the previous chapter, the scale of nanotechnology and fundamental biological processes are often one in the same. This means that nanotechnology provides potential tools for probing and manipulating biological processes. With enough understanding, it is expected that technology can be devised to both monitor and to tinker with any bodily function in a precise and controlled manner—this is the promise of biotechnology.[8]

Among the promised advances borne out of applying molecular chemistry to biology are improved drug-delivery systems. Analysis of the cellular structures has revealed a host of identifiable molecular-scale characteristics that may be used as markers for identifying cells for diagnostic and treatment purposes. Medical equivalents to smart bombs may be developed that use specific molecular markers as both "guidance" and triggers.[9] Only in the presence of targeted cells would a drug activate, meaning that other cells would be left alone. This is of particular interest as a means of delivering cancer-fighting drugs only to targeted cells. By targeting specific cells, such as those in a tumor, healthy cells are not exposed mitigating the side effects of the treatment attacking healthy cells. In general, this approach can be applied to many biological functions to produce drugs with reduced side effects.

Borrowing again from the military lexicon would be various biotechnologies that aim to camouflage drug molecules from the body's defenses. Large doses of many drugs have to be applied, again with possible side effects, due to the body naturally breaking down or outright attacking the treatment itself. There are also parts of the body that have natural barriers, such as the blood-brain barrier, that have limited the options for pharmaceutical treatment by simply being inaccessible to most drugs. Just as stealth and precision-guided weapons have proven to be force multipliers for the United States, the combination of targeted drugs with more survivable drugs, in the face of the body's defenses will allow for old medical strategies to make use of smaller dosages with less collateral damage, as well as opening new possibilities for medicine.

Finally there is the prospect of using technology to engineer biological processes to behave in ways not intended by nature. This is the transition from basic science, the understanding of how nature works, to controlling it for use in medical and even industrial procedures. Regenerative medicine is just such an example with a direct link to military use. Defense Advanced Research Projects Agency (DARPA)[10] and other military agencies[11] are funding research into various methods tissues, organs, and perhaps even limbs that can be coaxed into regenerating. Along similar lines, and possibly of nearer term fruition, is blood pharming, the in vitro (out of body) growing of red blood cells to ease the many supply challenges faced by battlefield medics.[12] Biotechnology goes beyond improving military medical care. The aging population of the West is clearly a market for products that maintain and improve quality of life. Futurists go as far as predicting that biotechnology will be able to first understand and control the processes that limit lifespan.

Improved emergency medicine and long-term care for veterans are not only expected, but should also be demanded, as part of how any sensible society chooses to treat its protectors. This still leaves the question of what else the military will to do with the opportunities made available by biotechnology. In the near term much of what is being researched by DARPA, the perennial target of commentators looking for scary new military technologies, and other military research organizations falls under the mundane, the routine activities that together make up the necessary foundations of a modern 21st-century military. These activities range from programs to develop better training regimes,[13] augmentation of a soldier's ability to absorb information in the midst of a battlefield,[14] and exoskeletons to assist with logistical tasks.[15] Essentially, biotechnology will start off as a force multiplier in carrying out familiar and somewhat mundane tasks. However, as the military does have a reputation, indeed an institutional responsibility to find new weapons, some are fearful that what starts off as better medicine will turn into something much more terrifying.

Keeping the Soldier Going: Metabolic Sustainment and Dominance

Food is an enduring defense biotechnology product from well before the time when the term technology was in use. Though some would argue against labeling military field rations as food, the ability to keep soldiers, sailors, and air crew nourished and in good health is ultimately the key factor to a nation's ability to field persistent global influence. Research by Royal Navy surgeon James Lind led to a nutritional solution, lemon or lime juice,[16] to the problem of scurvy in long-duration sea deployments—contributing to the Royal Navy's success at sea in the 19th century. The amount of time a U.S. nuclear submarine can remain underwater is constrained to a large degree by the amount of food that can be packed aboard. Perhaps underscoring both the rise of the private military contractor, and the ancient link between food and morale, is the phenomenon of finding U.S. fast food chains setting up shop in the middle of U.S. bases in Afghanistan and Iraq.

The need for a more compact, transportable, and easily consumed ration has been accelerated by combat experience. Research into how soldiers repackaged the meal, ready-to-eat (MRE), which initially entered service at the end of the Cold War, found that much of the nutritional content of each MRE was discarded as infantry prepared to leave base.[17] Throwing away part of the standard MRE for easier carriage was both wasteful, MREs are expensive, and unhealthy as the discarded food is needed to both fuel the body during high-intensity activities such as combat operations and also to aid in its recovery afterward. This research led to the development of the First Strike Ration© (FSR™) which is 50 percent smaller in volume and weight compared to the MRE,[18] but still contains a minimum of calories and other nutrients needed for combat operations.[19] This ration is under continual development to fit in additional dietary supplements, and expand the menu.

Beyond basic nutrition is the military equivalent to the sports drinks used to enhance athletic performance, and the energy drinks used by students to help with concentration and mental activity. Prolonged physical exertion will over time start destroying muscle tissue and damaging other organs. Added with stress of combat, it becomes clear why extended combat operations have a debilitating effect on soldiers.[20] Many of the compounds now being introduced into combat rations are meant to counteract tissue damage, quicken recovery times, and boost the immune system in response to combat stresses.[21] The term nutraceutical has come into use to describe the combination of food with medicine. DARPA as part of its 50th-year anniversary highlighted its own involvement in the development of these nutraceuticals that are beginning to appear in soldiers rations.[22] Another term for foods that do more than fulfill nutrients requirements is "functional food." The umbrella term used by the Department of Defense for these additives is performance-optimizing ration components (PORC). Indeed many of the items already being used to fortify foods or are under consideration for future inclusion, such as caffeine, vitamins, antioxidants,[23] and probiotics,[24] are things highlighted in the marketing for dietary supplements and energy foods. According to U.S. government numbers, 75 percent of the U.S. population takes dietary supplements[25] forming a market worth over $20 billion dollars a year.[26] The combination of people seeking longer and healthier lives, along with recreational and competitive athletes looking for a (legal) advantage is fueling growth in this area of biotech.

Requirements specific to military logistics (long-term storage without refrigeration, delivery to warzones) mean that military labs, such as those of the US Army Natick Soldier Research, Development & Engineering Center, are also active participants in food developments. Specific operational environments and needs have driven the development of meal, cold weather and the food packet, long-range patrol (MCW/LRP), as well as foods that seem to sacrifice resembling food to fit adequate nutrition into the most compact form possible for use in survival rations found. Additionally, unlike health food fads, the effects of specific food components have to be proven along with effective dosages before they can become a procurement item. One such study, out of Canada, linked specific dosages of caffeine with performance metrics of several infantry duties, including marksmanship.[27] Of note is that the study introduced caffeine directly to the bloodstream via chewing gum that released caffeine for absorption by the chewer's gums.

A potential biotechnology product expected in the long term related to nutraceuticals are edible vaccines produced in genetically engineered plants. It has been proposed that genetically modified (GM) food plants could produce specific proteins that would train the body to fight specific pathogens. Unlike traditional vaccines these proteins would not be sterilized pathogens, but just surface components of the targeted virus or bacteria,[28] enough for the body's immune system to recognize the real thing. Although much of the research is aimed at edible vaccines as a low-cost way to vaccinate people in the developing world,[29] it would also ease the logistical burden of preparing soldiers for deployment in these very same parts of the world. Global reach and persistence exposes Western military forces to many of the same challenges faced by those attempting to project humanitarian aid globally. With the trend of Western military forces being used increasingly for humanitarian interventions, edible vaccines would be a potentially low cost but very useful tool in the arsenal for soft-power missions.

Like many new technologies, or complex topics, and edible vaccines ranks highly in both categories, there is potential for backlash. In the West there is the persistent and unfounded panic over vaccinations.[30] Even the military is not immune to these fears, resulting in studies investigating such claims.[31] GM foodstuff in food aid has already faced some criticism from the developing world.[32] Now the biotech industry is not entirely blameless for this situation; scandals involving shoddy non-GM products, and missed opportunities for communicating the realities of GM food to the general public[33] have led to resistance toward GM foods and biotech in general. Miscommunications, media scaremongering, and outright ignorance of the scientific realities have been harnessed by everyone from political action groups against mainstream medicine and the biotech industries, to Islamic hardliners[34] trying to portray Western charity in a bad light.

Similar to the direct intake of caffeine into the bloodstream via the blood vessels in the mouth would be the transdermal nutrient-delivery system (TDNDS). The skin, being a very effective barrier against infection and other hazards of the world, is not meant to be an entry point for nutrients, but is permeable to various chemical and biological processes. The nicotine patch, used to help wean smokers off cigarettes, is probably the most visible application for transdermal delivery today. A long-term vision, the TDNDS hit the headlines prior to the Global War on Terror, presenting more immediate feeding R&D needs.[35] Part medical lab and part nutrient-delivery system that bypasses the digestive system, TDNDS would actively monitor a soldier's health during combat and deliver nutrients and nutraceuticals into the body via advance forms of the nicotine patch. It would theoretically react to counter the physiological effects of hunger, exhaustion, stress, and injury by varying the mix it was supplying to the bloodstream.

At the time, it was unclear if this technology was even viable,[36] though it does have potential linkages with other emerging technologies—small medical monitors that are worn, ingested, or implanted, along with the computer software needed to react to bodily needs figure highly in advance computing and nanotechnology categories. Similarly the delivery mechanisms could include microelectromechanical systems (MEMS) to release nutrients and medicinal compounds directly into the bloodstream. Protecting nutrient and drug molecules in nanotechnology shells that only release their payload once reaching a targeted part of the body is an emerging technology that can also be applied to the TDNDS concept. Research continues on noninvasive transdermal-delivery technology, some of it funded by the military, such as at MIT's Institute for Soldier Nanotechnologies.[37]

Probably not much more controversial than having soldiers absorb food through their skin is the idea of using the same technology to deliver therapeutic medicines. In addition to nicotine delivery, other medical therapies are already delivered via transdermal patches, such as painkiller, heart medication, and mood stabilizers. In the context of future infantry gear this delivery method would be used in conjunction with active health monitoring and active control—although the transdermal-delivery system is part of the uniform, and hence in contact with the application area of skin, it would only deliver medications on command by either remote control from military medics, or by built-in medical computers that will autonomously attempt to keep a soldier optimized for combat. Remote or autonomous therapeutic medical attention is expected to continue the reductions in treatment times that have increased battlefield survival rates. However, with the possibility of treatment comes the possibility of enhancement; this is where some become uneasy at the prospect of a drug-enhanced fighting force. Biotechnology is only providing the tools for these efforts to enhance soldier performance; the use of specific enhancements is a policy question, though one made possible by the state of the art in medicine.

Successive U.S. Army programs, Objective Force Warrior, Future Warrior, and the current Future Force Warrior have to varying degrees offered such health monitoring to enhance soldier performance. In its current guise, this is the warfighter physiological status monitoring (WPSM) system found in discussions on the Future Force Warrior program.[38] Many long-term plans for future soldier uniforms/personal combat vehicles (exoskeletons) feature materials that can harden on command, alternatively providing armor protection or acting as ad hoc medical devices to immobilize injured limbs and apply pressure as a tourniquet.[39] The smart materials needed for a soft uniform to convert on command into hard bullet-resistant armor or immobilize a limb only when needed are probably further in the future than automatic and remotely controlled drug delivery. Embedded sensors in soft uniform materials able to network with other medical systems are simply an application of the smart materials now entering the market. Programmable drug-delivery systems already exist in the form of various implantable insulin pumps, though this technology has logistical, survivability, and likely end-user resistance problems. For sheer practicality, noninvasive technologies such as transdermal delivery are much more likely to be adopted than subjecting certain military occupations to mandatory surgery.

Related to integrating medical monitoring and autonomous first aid systems into combat uniforms is the deployment of teleoperated surgical robots to the battlefield. The U.S. Army's Telemedicine and Advanced Technology Research Center (TATRC) and DARPA have been working on programs such as Trauma Pod, which aim to produce an unmanned, mobile operating room.[40] Although there are many challenges that remain for totally autonomous medical care, telemedicine promises to link doctors located far from the frontlines to those who need medical care. Remote-control triage care, along with the various robot programs geared toward extracting the wounded,[41] and possibly start treatment during extraction,[42] will cut down on response times, potentially increasing survivability on the battlefield.

The flip side of providing soldiers with the energy needed to endure for long periods in combat would be dealing with all the waste products generated by metabolism. An example would be the by-products and side effects of muscles operating in an inefficient anaerobic, without oxygen, mode due to working at levels that exceed the body's ability to deliver oxygen.[43] The pain felt by n body overtaxed by strenuous exercise is a warning sign that the body has been operating in this mode. Extending this envelope where the body is able to avoid anaerobic muscle operation is of interest to both the sporting and military communities—as would be methods to enhance the body's ability to deal with excess metabolic wastes and by-products of exertion. The former would increase endurance as it would train the body to work in a more sustainable manner. The latter would increase the amount of time a body could sustain intense activities, such as sprinting, as well as shorten recovery times after periods of such activity.

Along with metabolic waste, the body also produces debilitating amounts of heat during physical exertion. Sweating is the natural coping mechanism for this, but this has side effects such as dehydration, and is not always an option as soldiers become more enclosed in protective gear. Using technology to keep soldiers cool is in some ways not much of a future capability—body temperature regulating suits have been a requirement for spacesuits since the early space age. However, much of the research linking heat, fatigue, and performance is quite recent,[44] and the goals are different: in the space program the purpose of personal climate control was to protect from the hostile environment of space; for the soldier it is sustain combat operations and increase performance by easing the load on the body's ability to regulate heat.

Many future infantry programs are including active temperature regulation as a means to increase endurance. Externally powered micro climate-control suits have been developed for pilots, and more recently for vehicle crews on patrol, to extend the amount of time for which they are effective in hot environments, such as Iraq.[45] New textiles with integrated active elements such as heat-conductive fibers and smart materials engineered to help transfer away heat are certainly options to producing less bulky and mass-deployable personal climate-control systems than those now in use for the space program.

The Mind of the Future Combatant

Though it may be a cliché, the mind is perhaps the key to enhanced performance overall. Just as the Information Age has flooded the workplace with e-mails and memos, the Information Age battlefield has its own forms of information saturation, along with e-mail and memo traffic. Concerns about information overload go beyond productivity metrics for the soldier; the ability to manage battlefield information may have an impact on personal survivability and mission success. Better-designed technology, expert systems, and forms of artificial intelligence will streamline how information is presented, but in the near and medium term it will still be the human mind that will be the limiting factor. Although one does not have to subscribe to decision-making models such as Colonel John Boyd's OODA (observe, orient, decide, act) loop, no one would desire a slower or impaired thought process. This research is now allowing biotechnology to be entering the market that promises to augment natural decision making and other thought processes, essentially technology to allow soldiers to outthink the enemy.

Caffeine can be described as a socially acceptable stimulant, and is perhaps only the tip of the iceberg, and where biotechnology starts entering the controversial territory of adjusting the state of a soldier's mind. Practical needs in the middle of combat have in past, and will continue to in future, required soldiers to remain functional for inhuman lengths of time. Training, preparation, and caffeine can only do so much and there remains a need to use controversial drugs to keep soldiers in fighting condition for lengths beyond natural human endurance. Militaries have historically turned to stronger pharmaceuticals than caffeine to push back exhaustion and keep soldiers fighting longer. This included the use of amphetamines by all sides during World War II.[46] Medical authorities in United States and other militaries still find utility in stimulants such as dextroamphetamine under very specific circumstances where sleep deprivation is unavoidable, and under medical supervision,[47] though it is noted that due to side effects, and overall cost to medical infrastructure, pharmaceutical options should not be the first choice.[48]

Although operation needs may force the use of stimulants, the risk should be appreciated. Long term, many stimulants, amphetamines, and related drugs have a potential to form addictions. In the short term side effects include behavioral and mood changes. As a neurologically active substance there is a risk of impairment to the senses and overall thinking process. It was alleged by the defense during the court martial of U.S. Air National Guard Major Harry Schmidt that side effects of command-approved "pep pills" contributed to the friendly-fire incident.[49] Used incorrectly, the impairment brought on by lack of sleep is replaced by impairment brought on by the stimulants. Additionally, overdoses can lead to death through a variety of mechanisms, and over time physiological effects may cause long-term health problems. Eventually the body will require sleep and rest to recuperate—something medication will probably never change.

The research into how the human body deteriorates between periods of rest as well as the technology to counteract such physiological and psychological breakdowns are dual use in nature. Medical research is ongoing into studying both the general nature of sleep as well as specific sleep-related disorders. Better treatments for sleep disorders are to be expected but there is also a market for over-the-counter prescription, as well as illicit energy boosters and stimulants. The popularity of energy drinks is indicative of this market, as well as its dangers. There are now concerns over the health effects of energy drinks, both in long-term consumption as well as their potential for an "overdose." The nutraceutical qualities of these products also introduce the problem of how they interact with other drugs and alcohol. Military-funded research could be useful for this market in that it provides objective standards for effectiveness, as well as a second look at safety.

Building on research into methods to fight off fatigue and the need for sleep is research into enhancing cognitive performance. Warfare in the 21st century is as much about information as it is about brute strength, which means that the more effective soldier is one able to process and react to information faster. Outside of combat, warfighters are required to master a great amount of technical information concerning equipment and operations. Although computers may be able to perform number-crunching feats such as discerning hidden patterns from the every growing intelligence databases, a human analyst will still need to put a context to what is found. These are the national security factors that have fuelled research into not only how the mind works, but also into how to improve it.

National security is not the only driver for tools to improve the mind; the civilian world in general has become more competitive, dispelling any notions of a laid-back and relaxed technological-assisted 21st century. The U.S. National Research Council 2008 report, *Emerging Cognitive Neuroscience and Related Technologies*, noted that market pressures from both an aging society and from healthy individuals seeking a competitive advantage would be a great driver in the development of cognitive neuroscience technology goods and services. The report also noted that the use of these drugs for enhancement was largely underground,[50] in that these drugs are meant as therapies for failing and abnormal cognitive abilities, attention-deficit hyperactivity disorder (ADHD) for instance, but instead are being used by healthy individuals to boost memory and ability to learn. It should be noted that many of these drugs are controlled substances and, among other side effects, are potentially addictive.[51]

The prospect of biotechnology being able to create drugs specifically to boost brain performance, even with reduced side effects, is for some a worrying idea as it can be viewed as the intellectual equivalent of steroids in sports. What may be illegal in sports may simply be an open industry secret in business, something not talked about but something that everyone does. There is, however, an emerging discussion on the ethics of such practices, and as cognitive enhancement is often rolled into other debates on the merits on transhumanism.[52] Wide-scale military use could inadvertently legitimatize such activities. Like other stimulants, rigidly controlled military use of cognitive enhancers may actually be less controversial than the prospect of such usage becoming commonplace in the civilian world.

It cannot be overstated that military service exposes the warfighter to extreme amounts of stress. Ongoing research is both confirming the linkage between stress and less-than-optimal performance as well as finding ways to deal with stress to improve performance. Additionally, psychological stress after deployment is presenting costs to both individuals as well as military and veteran health care. A 2012 Congressional Budget Office (CBO) report high-lighted the significant cost difference in treating recent veterans with post-traumatic stress disorder (PTSD) and those without—$8,300 with PTSD (to $13,800 when combined with traumatic brain injury (TBI)) versus $2,400 for those not diagnosed with PTSD.[53] Now these are average costs; individu-als react to stress differently, meaning susceptibility to and severity of PTSD, and associated treatment costs will be different. Personal costs, however, are often incalculable.

Technology continues to increase survivability for bodies, and with greater understanding of how the body and mind are connected, technology will be used to help protect the mind. The identification of TBI is one such advance. The combination of protective gear that allows survival with reduced bodily injury and enemies that favor the use of explosive traps has led to a seeming increase in neurological injury.[54] This has led to a redesign of combat helmets that better insulate the brain from shock. On the diagnostic side there has been growing attention to TBI both from the military experience, as well as from the world of professional sports, where the long-term effects of hard hits to the head are now being understood. TBI has been linked to decreased cognitive performance, as well as behavioral changes that may haunt a soldier years after service. Although it would seem the more that is discovered, the more vulnerable people appear, the outcome of better understanding injury mechanisms is leading to greater levels of protection against what were previ-ously unknown threats.

Training regimes, backed up by technology borne out of cognitive re-search, have the potential to reduce the amount of stress on the battlefield, and reduce the amount of psychological trauma afterward. Virtual-reality simulations, in addition to providing effective, safe, convenient, and low-cost training opportunities are also providing a window into the workings of a soldier's mind under many of the same stresses faced in combat. Programs such as the Stress Resilience in Virtual Environments (STRIVE) Project are combining simulated combat and noncombat situations with active mon-itoring of stress and other emotional markers to help prepare soldiers for deployment.[55]

The reality of military service is that there is a potential that one will be called on to take another person's life. Though such acts are necessary for the protection of others, and indeed securing entire societies, it is still for the av-erage person a traumatic act. Scholars such as retired lieutenant colonel Dave Grossman and others have written extensively on the psychological costs of killing. In the context of emerging military technologies, this is becoming a concern with drone operators who, thanks to advanced optical sensors, may have an up-close view of their actions and the results despite being physically located a continent away.[56] At the same time, many in civilian society find it very disturbing that there are training regimes designed for the purpose of preparing young adults for the act of fighting and killing. It is therefore of no surprise that the use of biotech products to assist in this preparation would be controversial.[57]

To be clear there is little need in a modern professional military for uncontrolled violence. In this media- and information-saturated era, the lack of judgment and outright crimes of a very small number of uniformed personnel have had a magnified effect, leading to unjust charges against all past and current serving personnel.[58] Additionally, these sensationalized incidents are often being twisted by anti-Western forces as propaganda and justification for terrorist attacks. Excess use of force is clearly something to be avoided, not just on humanitarian grounds, but for purely practical reasons. Overall it must be remembered that the current Western paradigm for military action has been on increasing control over destructive force. Therefore it would seem more likely that military leadership would desire a greater control over aggressive behavior in military personnel than any kind of enhancement of aggression. Perhaps it is better to say that what is wanted is not increased aggression, but more on-demand control over the emotional states and mental stress while in combat. Refined knowledge of the neurotransmitters, hormones, and other brain activity in stressful situations is being applied to produce a calmer more effective soldier.

The human brain is perhaps the most complex organ known, but ongoing research is slowly revealing its secrets. A national security and law enforcement spin-on from the existing medical and recreational demand for portable and low-cost brain-activity monitoring is interest in using this technology as a lie detection or interrogation tool. With international terrorism remaining an ongoing threat to the West, the threats that brain-scanning technology is meant to counter are clearly known. Many intelligence and law-enforcement tasks revolve around determining whether a subject is relaying information accurately and techniques for probing the mind, noninvasively, seem like an ideal capability to develop.

Technology is providing a better picture of how mind and body work together. Sensors of increased resolution are linking thought, behavior, and actions with brainwaves, measures of brain's electrical field by electroencephalography (EEG), blood flow by functional magnetic resonance imaging (fMRI), and other measures. There is enough resolution to allow brain activity to maneuver wheelchairs and operate prosthetic limbs, though not to the level where thoughts and words can be captured. This research is also being refined for use as a means of lie detection. It is hoped that the neurological activity involved with lying, and other responses useful during an interrogation, would provide a better signature than body language and other physiological responses.

Among the techniques that have been studied by law enforcement and military agencies is brain fingerprinting. This technique looks for the presence of brain-wave patterns associated with recognition, or as one of leading proponents and developers Dr. Lawrence A. Farwell puts it, an "aha!" response.[59] Not specifically a lie detector, brain fingerprinting, through the combination of an EEG and carefully designed line of questioning, purports to be able to detect the presence of specific information. For example, brain fingerprinting could use recognition of the phrase, "No, not the mind probe," to separate out classic *Doctor Who* fans at a science fiction convention. A more practical use would be to quickly separate terrorist suspects from the innocent on the basis of whether or not they recognized specific knowledge of terrorist activities, personnel, and paraphernalia. Now as this is simply a test of recognition; investigators will still have to be mindful of context—an innocent relative of a known terrorist would have a strong recognition response to stimulus such as pictures of their dangerous relation. It is hoped by their backers that brain fingerprinting and related lie-detection techniques based on neurological observation will become trusted investigative and judicial tools.

Going further is the more controversial subject of using biometrics to scan a population for signs of terrorist and criminal activity before the actual attack or crime. Though the concept itself is criticized as being part of the Big Brother state, the Department of Homeland Security has had several programs aimed at identifying not just individuals, but also intent based on recognizing emotional and behavioral cues.[60] This is similar to data mining, except instead of looking for precursor transactions to terrorist or criminal acts in a person's digital paper trail, looks for subtle behavioral signs that are theorized to be precursors to terrorist or criminal acts in the physical world. As of 2011, testing in the field has begun. Opinions vary on these programs as either promising security tools, massive invasions of privacy, misspent funding,[61] or some combination of the three.

The prospect of a being able to recognize intent, concealment, and even guilt through these new technologies raises many ethical questions. A major question is effectiveness, how reliable must the technology be before it is employed. If the reliability rate of detecting evil intent is insufficient it can at best be only used as a screen for other methods. In the Global War on Terror, there have been several incidents of detention and allegations of torture of people who were later proved to be innocent.[62] At the same time there is the worry that terrorists may be able to slip through the system, again for reasons of human or systemic error. Backers of neurological, cognitive, and behavioral techniques claim that these technologies will mitigate the occurrences of both unpleasant cases.[63]

Related is the question of how far may pressure tactics be permitted if concealment is detected. Parts of biotech are geared toward more individualized medicine; it is conceivable that branches of biotech could be applied to individualizing interrogation. As former secretary for defense Donald Rumsfeld has proven, some people regard standing for hours on end as part of a healthy day at work, whereas for others it is regarded as torture. In practical terms these technologies could allow for a systematic and unbiased separation of detainees who respond to strenuous questioning, and those who require additional stress. This in theory would protect some from unnecessary pressure being applied, but then raises the question of what to do in particularly hard cases. These are polarizing questions, and the introduction of new interrogation technology, no matter how accurate, will not make them go away.

Genetics

The study of DNA, genetics, and the way the instructions for all life is encoded has moved on from the generally benign world of discovery, to the more controversial world of application. An organism's genes may be studied to determine its traits and characteristics, including its susceptibility to disease, environmental conditions, and toxins. Though the costs still remain high, and our knowledge of what each gene does is still limited,[64] there are fears that the application of genetic technology will bring about genetic discrimination, where opportunities may be barred to individuals on the basis of their DNA. Instead of screening on the basis of medical health, merit, and integrity, all things that an individual has an influence over, there is concern that medical insurance and other aspects of life could be screened based on genetic probability. As the military is an organization that must screen its personnel for fitness and suitability for membership, the question of whether genetics should be included in recruitment screening will be a possibility as DNA analysis technology becomes more widespread.

Control over genetics allows the mixing of genes from one organism into another, allowing one animal to express traits that it would never do so in nature. Genes are ultimately chains of molecules, chemistry in other words, meaning that with enough control over the formation of specific molecules, completely artificial genes can be constructed to achieve desired results. Not only does this allow the generation of life forms from just genetic data, but also potentially the creation of designer life forms. The molecular-level control needed to construct artificial DNA should be familiar as this is nanotechnology from the previous chapter in all but name. Genetic manipulation leads directly to biotechnology's capacity to produce weapons of mass destruction (WMDs). Even without its weaponization potential, there is debate over how to use the growing mastery over genetics; as in the genome, the complete set of genetic material for an organism is both the definition of its existence and instructions for its destiny.

Genetics: Biowarfare Potential

Biological weapons include both living pathogens, such as bacteria, germs, and virus, and toxins generated by living organisms, such as the ricin used in the infamous 1978 assassination of Bulgarian dissident Georgi Markov. Their utility in warfare has been very controversial; some regard them as terror weapons of limited tactical or strategic value due to their relatively slow and unpredictable effects compared to traditional kinetic options. Others have come to fear them as the "poor man's nuclear bomb," leading to its inclusion as a WMD. Without killing, illness is enough for pathogens to have debilitating tactical and strategic effects. There is great fear that commercial air travel and a generally more interconnected world may allow released infectious biowarfare agents to spread globally with catastrophic results. The limited spread and casualty numbers of pandemics such as SARS and H1N1 influenzas on the other hand indicate that the reality of a biowarfare attack or accident may not be as simple as the collapse of Western civilization. Indeed the limited casualty rate of the 2001 anthrax attacks against the offices of U.S. politicians and media would tend to indicate the difficulty in employing biological weapons. At the same time these incidents did generate a lot of panic, disrupted operations of these offices as well of the U.S. Postal Service, which was used unwittingly as the delivery mechanism, and did kill five U.S. citizens and sickened many others.[65] Perhaps beneficially these outbreaks and incidents have brought attention to the threat of infectious pandemics, both natural and manmade.

Befitting the label "poor man's nuclear bomb," biological and chemical weapons (toxins that are not necessarily from biological sources) figure prominently in asymmetric-warfare threats. Chemical and biological weapons are among the unconventional ways it is feared that a significantly mismatched opponent may choose to strike at the United States. Considering the U.S. strength in conventional and nuclear weapons, a rogue nation, by definition ignoring international norms, may resort to combination of state-sponsored terrorism with what mass-casualty weapons are available to it. With the expertise and equipment associated with chemical and biological weapons becoming more accessible, independent terrorist groups may choose these options as well. The Japanese Aum Shinrikyo doomsday cult used sarin nerve gas in several deadly attacks, culminating in 1995 with coordinated attacks of the Tokyo subway. It is known that Aum Shinrikyo also had invested in an unsuccessful biological weapons program.[66] Biotechnology has moved on in the last 20 years, and the challenges that stopped Aum Shinrikyo may no longer be relevant today.

There are two kinds of dual use with regard to modern biological weapons: research and development toward biological defenses can be easily converted into offensive biological weapons,[67] and the dual-use nature of the biotech industry itself. To develop vaccines and understand the threat posed by specific biological agents it is necessary to study that pathogen in detail. This includes information on how to culture a pathogen, and how to process it into a weapon such as an aerosol or particle suitable for dispersal over a wide area. The very same skill base and industrial equipment may be found creating large quantities of biological weapons agents, vaccines against the former, as well as medical products in general. With technology allowing the construction of smaller labs there is increasing potential for small undetectable biological weapons labs.

In addition to production, many of the biotech opportunities discussed earlier, such as targeted medicines and technologies to help deliver drugs past the body's defenses can be applied to biowarfare compounds to make them more lethal and importantly, in light of their questionable utility, more effective as weapons. Genetic engineering and synthetic biology can insert genes into pathogens to increase their survivability against both the immune system, as well as the environmental conditions that make biowarfare unpredictable. Toxins can be coated in molecular shells that evade the body's natural filters to deliver their payload to specific parts of the body.

In 2005, the Centers for Disease Control and Prevention (CDC) announced it had reconstructed the influenza virus strain responsible for the 1918 pandemic as part of efforts to understand this and related threats such as current bird and swine flus.[68] The techniques used by the CDC and others pioneering synthetic genomes[69] could also be harnessed for the purpose of bringing back other past killers, such as small pox, which have potential as bioweapons. It is of note that since the eradication of small pox, the CDC and its Russian counterpart both hold onto the last samples of small pox for, among other reasons, use as a safeguard against the event that the eradication of this killer disease was declared prematurely. At the time there were fears that these small pox samples could be used in biological weapons programs. Now with synthetic biology becoming a refined mainstream tool of biology and biotech, all that is needed to resurrect the virus is data on its genome.

Added to the mix are the beginnings of open-source biotech and biohacking, modeled after open-source equivalents in the computer field, as a means to encourage new discoveries and creative uses for genetics without the limits of restrictive copyright patents, and indeed to promote the free exchange of biotech components (genetic components). The international genetically engineered machine (iGEM) competition, started at MIT in 2005, has undergraduates from around the world to assemble machines out of biological parts, which are then inserted into living cells.[70] That undergraduates, high school students, and the general public are taking an active interest in synthetic biology skill set, is both inspirational and frightening. Though at the moment somewhat overstated, many of the tools and skills available to hobbyists may have applications in engineering biological weapons. Indeed if the parallels to tinkering with computers are carried further, harmless fun with the *E. coli* bacteria[71] will evolve overtime into rampant and widespread availability of malicious biological organisms and compounds created in underground labs staffed by brilliant, but unethical, enthusiasts. It remains to be seen what kind of safeguards will develop as more people take an active interest in tinkering with the instruction set of life itself.

The balance between biotechnology as a defense against biological weapons and as an enabler for biological attack is similarly unclear. The present prevailing notion is that the life sciences should be used to improve health and quality of life—increasing the likelihood that biotechnology will be overall beneficial, though some debate exists as to what definitions of beneficial apply. However, as history has shown that is not a universal belief, nor is it a straightforward one. Putting aside the dual-use nature of biological weapons research, at certain points in history the United States and other nations did believe that biological weapons had deterrent value, meaning that work on these most abhorred of weapons was linked to maintaining peace. As the threat of biological weapons cannot be wished away, it is expected that biotechnology will simply continue to be both a tool to mitigate the threat, and be under suspicion of enhancing it.

Genetics: Shifts in Global Production

Much less panic causing, but probably of more day-to-day relevance to the military is genetic engineering's potential to produce materials. Nature is already adept at using exponential growth to cover the earth in all manners of "manufactured" products. It must be remembered that without life producing them, many of the complex molecules that make up proteins and other cellular structures would simply not exist. All genetic engineering would be doing is inserting genes to cause organisms to produce molecules that would be useful for very specific human purposes.

Genetically engineered goats that produce spiders silk[72] gain attention in part due to the somewhat disturbing concept of making a large mammal produce something commonly thought of as only being produced by small insects. On closer analysis, the key components of spider silk and milk are both proteins, though ones with vastly different mechanical properties.

There are no natural reasons for spider silk protein to exist in milk; however, there are man-made ones. Artificial spider silk has long been suggested as a future material for high strength but lightweight applications such as body armor. The principal barriers to use are the challenges in bulk production. Biotechnology is in an odd way of potentially returning soft body armor to its roots. Early bullet-resistant vests were made of layers of silk. Real silk comes from silkworms, which started being harnessed for silk production centuries ago.

Of less surprise to the general public is biofuel production. Fossil fuels are by some definitions organic products; at one point in time their source materials were living plants and other organisms such as plankton. This dead organic matter after being trapped under layers of sediment for periods of time measured in eons is compressed until it forms fossil fuel, which one specifically depending on the starting materials and nature of the geological events that formed it. Given enough energy any organic material can be processed into substitutes for fossil fuels. An example of this is oil from the Canadian tar sands, which is a purely industrial process.

Biofuels are fuels produced generally from harvested plant matter, though feed stocks may also include dead animal matter. Biofuels include ethyl alcohol (ethanol), which is produced in the same method as alcoholic beverages by the fermentation process, but for compatibility with existing engines must be blended with traditional petroleum products. Seed oils, a familiar item in most kitchens, have also been processed into fuels, some of which have already been tested by the U.S. military. In 2011, a U.S. Navy trainer was test flown with a fuel mix that included a component produced from Camelina seed oil.[73]

Genetically engineered plants may provide both better yields of feed stocks, such as sugars and starches. This technology has also been used to produce strains of bacteria able to process organic waste materials such as cellulose.[74] Various labs in recent years have been able to modify bacteria to produce usable fuel directly,[75] in a process again resembling industrial fermentation; however, as it is chemically identical to traditionally produced fuels, it does not have to be blended like ethanol and other biofuels.

There is, however, a price to this—currently available biofuels are more expensive than fuels produced from traditional petroleum sources. Indeed most alternative sources of petroleum products, including tar sands, have a tendency to be more expensive than crude oil due to the need for more processing. At the time of writing there was an ongoing controversy over continued funding of U.S. military biofuel projects in light of severe funding constraints to other priorities.[76] This, however, may not always be the case as traditional petroleum prices are subject to large price fluctuations, as they did toward the end of the first decade of the 21st century. Though controversial, government-subsidized research on biofuels, including that from the military, may potentially lower prices, as research into directly using biotech to produce "drop-in" biofuels creates a new, and practically inexhaustible, source of pure fuels. Alternatively both demand and prices for fuel may simply remain high, allowing for biofuel costs to be lowered while still being a profitable investment.

Another political dimension to biofuels is energy independence for the United States. There is a perception that the United States obtains too much of its fuel by importing it from dangerous parts of the world—including linking doing business with the Middle East with indirect support for terrorism.[77] This is a complex issue, as even the definitions of energy independence and energy security are open to debate.[78] Some view increasing domestic supplies of oil as a bulwark against foreign pressure from oil-supplying nations. This has happened before, in the 1970s, when Middle Eastern nations put an embargo on oil exports to the United States over U.S. support of Israel. Use of oil as a blunt political tool would be difficult as all who participate in the petroleum market would be affected. Even the linkage of oil imports to instability in the Middle East and elsewhere is questionable. Increasing domestic production of petroleum would have little effect on the global price of oil,[79] and therefore little impact on oil revenues (accused of directly or indirectly being funneled toward terrorism).

Man and Machine: Workspaces and Mobile Infantry

While waiting for the advent of artificial intelligence, improvements to the interface between humans and machine will mitigate the limits of the slow piece of wetware still required in some form for all weapon systems. Even experienced computer users sometimes find themselves wishing that the computer could read their intent, instead of their actions, when trying to get a computer to perform some desired action. Often it is just human error that prevents people from getting the most out of the computer system, and other times it is the interface that is the barrier to getting the most out of technology. Although it is often said that people limit what weapon systems can accomplish, another limiting factor is the interface between operator and machine. Cognitive research is not just about drugs to improve mental alertness; it also includes branches that study just how people interpret information, and perhaps how the rate information is supplied can be optimized.

The military has been on the forefront of improving the linkages between man and machine. Decades ago the first heads-up-display (HUD) and hands-on-throttle-and-stick (HOTAS) were introduced into combat aircraft, allowing a pilot to conduct an attack without looking down at cockpit instruments and displays. These innovations only started making inroads into civil aviation recently, and are available in some higher-end automobiles. The computer revolution has allowed HUD systems to become small enough to be fit safely into helmets. In the civilian world, where overburdening the neck under high g-forces is not so much of a problem, wearable computer interfaces, such as Google's recently unveiled Project Glass, are promising to bring head-mounted displays (HMD) to the masses. Additionally, the somewhat maligned[80] Land Warrior program did deploy monocular (one eye) displays with soldiers sent to Iraq and Afghanistan, despite the program being largely defunded.[81] Prototypes now exist of displays reduced down to contact lenses as part of DARPA-funded research.[82]

The technology being worn has certainly made progress; however, what is of more importance is the information being displayed. Sometimes called synthetic vision systems, the current buzzword is augmented reality. Provided the display knows where it is and how it is oriented, it can overlay data over the seen image. For military users this will include imagery generated by night vision or thermal-imaging equipment. Aviators may have a complete unobstructed view in all directions (or at least in all directions there is sensor coverage) to allow for superior situational awareness.[83] As with all hi-tech military concepts, augmented reality is expected to be networked, meaning commanders monitoring the mission from headquarters may not only be able to see the same thing the a soldier on patrol is seeing, but also be able to mark for attention objects and people of interest found in the soldier's field of view. Indeed some application of cognitive research aims specifically to assist enhance how specific information is communicated.

Essentially with augmented reality, it is hoped that much of the thinking can be offloaded to a computer, or specialists in headquarters. Augmented reality, if used as the interface for earlier net-centric warfare concepts, could help integrate infantry[84] into the network of vehicle- and drone-based sensors, allowing for a target seen by one to be passed on to another even without direct line of sight. Another example would be autonomous language translation software generating subtitles that could be overlaid in a soldier's field of view. This would allow the soldier to maintain eye contact with the foreign speaker, understand what was being said despite having no knowledge of the language, and importantly provide feedback that would help break communication barriers. A rosy picture to be sure, but as long as winning hearts and minds remains a goal, technology that helps to build rapport with the locals will be as important as applications that allow a soldier to destroy targets that they cannot naturally see.

Part of the interface is supplying information; the other part is inputting information. With the rise of social networking this now goes beyond issuing commands to worn equipment and includes communicating with others.

Eye tracking is the direct opposite of augmented reality displays; instead of a computer outputting an overlay of important information, what the eyes are focused on is used as an input. In the many areas of cognitive science, eye tracking generates some of data points involved with research on learning, perception, and even wakefulness.[85] Commercially available eye tracking is available as a mouse/trackball substitute. The closest military equivalent in service right now would be head-tracking helmets used in some fighter aircraft to provide weapon cueing.

Other minimalist computer input options include gesture and voice recognition. Leveraging off both the consumer electronics industry, and the field of biometrics, both these technologies hope to be able to translate natural verbal and nonverbal communications into something a computer can understand. If cameras can recognize intent just from subtle movement and body language, then it is not much of a leap that software can be devised to recognize movement and gestures as commands. Years before camera-based systems for personal gaming consoles became common, there was speculation that sensors embedded in gloves could be used as an input for silent communications between networked soldiers.[86] A limited form of voice recognition is already available in the Eurofighter Typhoon,[87] and may be available in the U.S. F-35 Lightning II,[88] a multiservice aircraft still under development at the time of writing. The challenge is, however, in developing software able to accommodate the many variations in speech within the same language. Combat conditions may also impact accuracy, limiting their application to noncritical tasks.

As noted earlier, brain waves can be read with enough resolution to navigate wheelchairs. Like the eye tracking, there are very rudimentary brain-wave-reading computer interfaces, brain-computer interfaces (BCI), available to the public. Although the technology is still in its infancy, it is of interest as a means of controlling prosthetics. DARPA funding (over $70 million)[89][90] is behind the Revolutionizing Prosthetics 2009, which aims to produce a neurally controlled artificial arm. Instead of reading muscle contractions as in current myoelectric arms, DARPA and its partners hope to hook a robotic arm directly to the brain for control. Two years later this program was hitting headlines again with the Food and Drug Administration (FDA) announcing that it would be fast tracking the approval process for the fruits of this project, the implantable sensor to control the arm, it entered human testing.[91] For many military service has meant loss of limbs and, therefore, it is understandable why the military is involved with research and development into artificial limbs, as well as making it available to soldier-amputees first.[92]

Associated with a brain control of a prosthetic would be the prosthetic providing sensory feedback such as touch, pressure, heat, and with a high-enough "resolution" ability to feed information into the brain, texture. Although it is unclear if a machine will be able to totally replace flesh and bone, without sensory feedback the artificial arm would only be a modest improvement over current-day prosthetics. Though of very low resolution so far, and limited in what type of vision loss they correct, visual prosthesis has already entered human trials.[93]

Avoiding the regulatory difficulties of an implant would be the powered exoskeleton. Also called a wearable robot, the powered exoskeleton also avoids the many technical barriers and troubling ethical problems of artificial intelligence or remote-control warfare by putting the combatant directly in the middle of the fighting, but enhanced by robotics. Alternatively, this is the combination of sometimes unpredictable and/or cognitively weak humans with the maintenance and power requirements of walking robots. There is truth in both descriptions of present-day exoskeleton technology today. In the near term, real-world progress is being made on military exoskeletons for use in more-benign settings as logistical machinery, basically a walking forklift.

Military interest in the powered exoskeleton goes back decades as a means to amplify strength for logistical tasks. In the 1960s General Electric attempted to produce a powered exoskeleton, the Handiman, to magnify the user's strength by 25 times.[94] Funded by the military, among the tasks for this large apparatus was loading ordnance onto aircraft onboard aircraft carriers,[95] something which to this day still requires a small team of sailors to perform. As a form of human amplification, Handiman was meant to be controlled by simply replicating the movement of the limbs it was worn over. Weight, control, and power problems have, however, stopped Handiman and other powered exoskeleton projects from being viable until recently.

Today's exoskeletons are much smaller than Handiman, with more achievable goals of simply taking the burden of an oversized combat load off the soldier,[96] or allowing those suffering from physical weakness or loss of mobility due to paralysis to walk again.[97] Faster computers and smaller microelectronic sensors have solved many of the control and balance problems, much in the same way they have only recently solved the many challenges of walking robots. Both an exoskeleton used as a mobility device for the paralyzed and one that will allow a sleeping soldier to continue marching are simply walking robots that have space reserved for the operator's legs and torso. The two current prototypes that DARPA is funding, the XOS series by Sarcos/Raytheon and the Human Universal Load Carrier (HULC) by Berkley/Lockheed Martin, have agility requirements.[98] Despite seeming being in competition, both exoskeleton projects seem to be fulfilling different roles. The former is a tethered system that includes amplified arms and appears more geared toward logistical work. The later is self-contained but comprises only a pair of powered legs meant, as it name implies, to only assist with the carriage of heavier loads than the standard combat pack, implying a role in augmenting combat endurance, though it also could be used for logistical tasks.

Despite smaller more efficient electrical systems (and the loss of the heavy hydraulics found in early concepts like Handiman) battery life is still measured in hours. This even applies to Lockheed's HULC, where there is some talk of deployment for testing in warzones such as Afghanistan.[99] Advance batteries, fuel cells, artificial muscles, and replacing electrical actuators with chemically powered ones are all on the table as far as ways to extend the power supply for an exoskeleton with enough endurance for extended combat operations are concerned.[100] Otherwise the exoskeleton will remain tethered to a power supply, much as the XOS2 prototype is right now.

Limits: How Far Is a Society Willing to Go

There are fears as to how far military biotechnology research will go. The drug-enhanced "super soldier" is a common trope for some in the media wanting to portray the military as ruthless and reckless. Unlike athletics, the military would seem to have the option to go further than just better rations, training programs, and computer interfaces. If relatively safe and side-effect-free performance-enhancement treatments can be developed, for the military at least, their use would not be problematic from a fair play standpoint—the stated goal of all military R&D is to produce an unfair advantage. However, this is a very simplistic view; the use of biotech to enhance the body is somewhat more complex. Reining in the nightmare, and for some the dream, of the "super soldier" are the many practical challenges and ethical barriers.

The real-world history of pharmaceutical means of enhancing body and mind has had a very checkered past. Many of these existing drugs are important medical therapies, but also have found many illegal uses ranging from recreational drug abuse to cheating in competitive sports. Under medical supervision they can give hope, but in the wrong hands more often than not they carry legal and health risks. Even under medical supervision, therapeutic use of steroids, although life saving in one area of the body, may cause lasting damage somewhere else. For doctors and patients it is often a difficult choice of debilitating side effects or death. It is important to remember that many of the promises of molecular and other better-targeted forms of medicine are just that—promises. The human body is the most complex organism known, with many secrets enduring continued assault by the life sciences. Although it is likely that efforts to reduce side effects will be successful, it is unlikely that all side effects will be eliminated.

Again it is practical needs that will drive what forms of pharmaceutical and other augmentation of physical abilities will be used. Even with limited personal costs, the benefits may not be worth the financial costs to the system as they would simply end up being only marginal improvements over already effective physical training methods and enforcement of existing standards. If one chooses a military career, one probably has been made well aware of the physical-fitness standards for particular paths to career success. If biotech solutions may not be worthwhile for general issue, their use by individuals may carry the same repercussions from the organization as "cheating." Alternatively there may be pressures within the organization to turn a blind eye or encourage such behavior. In a contemporary real-world military context, sometimes seeking out a fair fight is a foolhardy act, other times a level playing field is important to honor, unit cohesion, the organization, and the mission of defending the country.

It is entirely conceivable that some biotech options for enhanced performance may be available in the civilian world but not available within the military supply chain for reasons of policy. During the wars in Iraq and Afghanistan some controversy arose over the use of privately bought body armor.[101] Early in those conflicts there were supply problems with body armor, leading some soldiers to believe that to increase their own chances of survivability they had to invest in their own protective gear. The Department of Defense's position on the other hand was that the unregulated body armor may have been of an inferior quality to the standard issue. Perhaps cognitive and other biotech enhancements will be future nonstandard pieces of "equipment" that will put soldiers in conflict with military leadership.

Ethical considerations over applying biotech to enhance the warfighter mirror the general debate over how far biotech and other emerging technologies should be used to directly change people and indeed societies.[102] Distribution of the benefits of biotechnology is a common ethical debate brought up by futurists trying to predict how emerging technologies will change the world. The question of how far to go does not exist in a vacuum, others are engaging in research and development in these same emerging industries. Under most theories on international relations, biotech should be used as a tool to enhance a nation's place in the world; however, not all theories would agree on using it as a tool to enhance hard power. Cooperation on more benign areas of biotech, and even select militarized ones, could have security payoffs. Again it is a matter of what works.

Notes

1. Martin van Creveld, *The Transformation of War* (New York: The Free Press, 1991), 222.

2. Masaki Imai, et al., "Experimental Adaptation of an Influenza H5 HA Confers Respiratory Droplet Transmission to a Reassortant H5 HA/H1N1 Virus in Ferrets,"

Nature, May 22, 2012, http://www.nature.com/nature/journal/vaop/ncurrent/full/nature10831.html.

3. Martin Enserink, "Grudgingly, Virologists Agree to Redact Details in Sensitive Flu Papers," *Science*, December 20, 2011, http://news.sciencemag.org/sciencein sider/2011/12/grudgingly-virologists-agree-to.html.

4. David Axe, "Semper Fly: Marines in Space." *Popular Science*, December 2006, http://www.popsci.com/military-aviation-space/article/2006–12/semper-fly-marines-space.

5. George Friedman and Meredith Friedman, *The Future of War: Power, Technology and American World Dominance in the Twenty-first Century* (New York: St. Martin's Griffin, 1998), 392–93.

6. National Science Foundation, "Edible Vaccinations," http://www.ornl.gov/sci/techresources/Human_Genome/home.shtml.

7. Ray Kurzweil, "How My Predictions Are Faring," October 2010, http://www.kurzweilai.net/predictions/download.php.

8. Michio Kaku, *The Physics of the Future* (New York: Double Day, 2011), 121–24.

9. "Medical 'Smart Bomb' Possibly the Key to Eradicating Cancer Cells," *Deakin Research*, July 2010, http://www.gsdm.com.au/newsletters/deakin/july10/7.html.

10. Department of Defense, *Defense Advanced Research Projects Agency Justification Book Volume 1 Research, Development, Test & Evaluation, Defense-Wide Fiscal Year* (FY) *2012 Budget Estimates*, www.darpa.mil/WorkArea/DownloadAsset.aspx?id=2400.

11. Telemedicine & Advanced Technology Research Center, "Regenerative Medicine: Creating the Future for Military Medicine," August 2009, http://www.tatrc.org/ports/regenMed/docs/TATRC_regen_med_report.pdf.

12. DARPA, "Blood Pharming," http://www.darpa.mil/Our_Work/DSO/Programs/Blood_Pharming.aspx.

13. DARPA, "Education Dominance," http://www.darpa.mil/Our_Work/DSO/Programs/Education_Dominance.aspx.

14. DARPA, "Overview of the DARPA Augmented Cognition Technical Integration Experiment," May 8, 2008, http://www.dtic.mil/cgi-bin/GetTRDoc?AD=ADA475406.

15. B.S. Richardson, editor. *Phase I Report: Darpa Exoskeleton Program*, Oak Ridge National Laboratory, 2004, http://www.ornl.gov/~webworks/cppr/y2004/rpt/118274.pdf.

16. Arthur Herman, *To Rule the Waves* (New York: Harper Collins, 2004), 260.

17. Christen N. McCluney, "'First-Strike Ration' Aims for Better Nutrition," *American Forces Press Service*, November 25, 2009, http://www.defense.gov/news/newsarticle.aspx?id=56854.

18. Department of Defense, *Operational Rations of the Department of Defense*, http://nsrdec.natick.army.mil/media/print/OP_Rations.pdf.

19. Ibid.

20. Department of Defense, *USMC FM 90–44/6–22.5 Combat Stress*, https://rdl.train.army.mil/catalog/view/100.ATSC/AF18AF6D-7DFB-4CA4-BE78-8D0A5F EDF993–1274316004566/6–22.5/TOC.HTM.

21. Department of Defense, "Performance Optimizing Ration Components," http://nsrdec.natick.army.mil/media/fact/food/perc.pdf.

22. Jonathan Beard, "DARPA's Bio-Revolution," *DARPA: 50 Years of Bridging the Gap,* http://www.darpa.mil/About/History/First_50_Years.aspx.

23. Department of Defense, "Performance Optimizing Ration Components," http://nsrdec.natick.army.mil/media/fact/food/perc.pdf.

24. Department of Defense, "Probiotics, Prebiotics, and Keif," http://nsrdec.natick.army.mil/media/fact/food/Probiotics.pdf.

25. National Institute of Standards and Technology, "Dietary Supplement Analysis Quality Assurance Program," http://www.nist.gov/mml/analytical/organic/dsqap.cfm.

26. Ibid.

27. Tom M. McLellan et al., "The Impact of Caffeine on Cognitive and Physical Performance and Marksmanship during Sustained Operations," *Canadian Military Journal,* Winter 2003–2004, http://www.journal.forces.gc.ca/vo4/no4/military-meds-eng.asp.

28. Sharon Guynup, "Seeds of a New Medicine," *Genome News Network,* July 28, 2000, http://www.genomenewsnetwork.org/articles/07_00/vaccines_trees.shtml.

29. http://www.nsf.gov/od/lpa/nsf50/nsfoutreach/htm/n50_z2/pages_z3/16_pg.htm.

30. Chris Mooney, "Why Does the Vaccine/Autism Controversy Live On?" *Discover,* May 6, 2009, http://discovermagazine.com/2009/jun/06-why-does-vaccine-autism-controversy-live-on/article_view?b_start:int=3&-C=.

31. Sandra I. Sulsky, John D. Grabenstein, and Rachel Gross Delbos, "Disability among U.S. Army Personnel Vaccinated Against Anthrax," *Journal of Occupational and Environmental Medicine,* Volume 46, Number 10, October 2004, http://www.anthrax.mil/documents/library/Anthrax2004.pdf.

32. Lusaka Olga Manda, "Controversy Rages over 'GM' Food Aid," *Africa Renewal,* Vol. 16 #4, February 2003, http://www.un.org/en/africarenewal/vol16no4/164food2.htm.

33. Richard A.L. Jones, *Soft Machines: Nanotechnology and Life* (New York: Oxford University Press), 5.

34. Gilbert Da Costa/Kano, "Setback for Nigeria's Polio Fighters," *Time,* October 25, 2007, http://www.time.com/time/health/article/0,8599,1675423,00.html.

35. Jim Garamone, "Patch May Deliver Nutrients to Future Warfighters," *American Forces Press Service,* February 28, 2000, http://www.defense.gov/news/newsarticle.aspx?id=44529.

36. Department of Defense, "Arming Soldiers with Nutrition," *The Warrior,* January-February 2000, http://www.natick.army.mil/about/pao/pubs/warrior/00/janfeb/patch.htm.

37. Massachusetts Institute of Technology Institute For Soldier Nanotechnologies, "Project 2.3.2: Non-Invasive Delivery and Sensing," http://web.mit.edu/isn/research/sra02/project02_03_02.html.

38. Philip Brandler, "The United States Army Future Force Warrior—An Integrated Human Centric System," http://ftp.rta.nato.int/public//PubFullText/RTO/MP/RTO-MP-HFM-124///$MP-HFM-124-KN.pdf.

39. Popular Science, *21st Century Soldier* (Time Inc., 2002), 127.

40. Jacob Rosen and Blake Hannaford, "Doc at a Distance," *IEEE Spectrum*, October 2006, http://spectrum.ieee.org/biomedical/devices/doc-at-a-distance.

41. Gary Gilbert, et al., "USAMRMC TATRC Combat Casualty Care and Combat Service Support Robotics Research & Technology Programs," http://www.tatrc.org/ports/robotics/docs/USAMRMC_CCC_CSS_Robotics.pdf.

42. P. W. Singer, *Wired for War* (New York: The Penguin Press, 2009), 112.

43. Stephen M. Roth, "Why Does Lactic Acid Build Up in Muscles? And Why Does it Cause Soreness?" *Scientific American*, January 23, 2006, http://www.scientificamerican.com/article.cfm?id=why-does-lactic-acid-buil.

44. Noah Shachtman, "Be More Than You Can Be," *Wired*, Issue 15.03—March 2007, http://www.wired.com/wired/archive/15.03/bemore_pr.html.

45. C. Todd Lopez, "PEO Soldier showcases gear at Pentagon," Army News Service, June 16, 2009, http://www.army.mil/article/22755/peo-soldier-showcases-gear-at-pentagon/?ref=news-arnews-title1.

46. Peter W. Singer, "How to Be All That You Can Be: A Look at the Pentagon's Five Step Plan for Making Iron Man Real," *The Brookings Institution*, May 2, 2008, http://www.brookings.edu/research/articles/2008/05/02-iron-man-singer.

47. Dr. John A. Caldwell, "Dextroamphetamine and Modafinil are Effective Countermeasures for Fatigue in the Operational Environment," http://ftp.rta.nato.int/public//PubFullText/RTO/MP/RTO-MP-HFM-124///MP-HFM-124-31.pdf.

48. Ibid.

49. Canadian Broadcasting Corporation, "Go-Pills, Bombs & Friendly Fire," November 17, 2004, http://www.cbc.ca/news/background/friendlyfire/gopills.html.

50. Committee on Military and Intelligence Methodology for Emergent Neurophysiological and Cognitive/Neural Research in the Next Two Decades, National Research Council, *Emerging Cognitive Neuroscience and Related Technologies* (Washington: The National Academy Press, 2008) http://www.nap.edu/catalog/12177.html.

51. Art Markman, "Are ADHD Drugs Smart Pills?" *Psychology Today*, September 6, 2011, http://www.psychologytoday.com/blog/ulterior-motives/201109/are-adhd-drugs-smart-pills.

52. Daniel McIntosh, "The Transhuman Security Dilemma," *Journal of Evolution & Technology* 21(2)—December 2010, http://jetpress.org/v21/mcintosh.htm.

53. Congressional Budget Office, "The Veterans Health Administration's Treatment of PTSD and Traumatic Brain Injury among Recent Combat Veterans," February 9, 2012, http://www.cbo.gov/publication/42969.

54. Katie Walter, "A New Application for a Weapons Code," *Science & Technology Review*, March 2010, https://str.llnl.gov/Mar10/pdfs/3.10.3.pdf.

55. Albert Rizzo, et al., "Virtual Reality Goes to War: A Brief Review of the Future of Military Behavioural Healthcare," *Journal of Clinical Psychology in Medical Settings* 2011, 18(2), 176–87.

56. Paul McLeary, Sharon Weinberger, and Angus Batey, "Drone Impact On Pace Of War Draws Scrutiny," *Aviation Week*, July 8, 2011, http://www.aviationweek.com/aw/generic/story_generic.jsp?channel=dti&id=news/dti/2011/07/01/

DT_07_01_2011_p40–337605.xml&headline=Drone%20Impact%20On%20 Pace%20Of%20War%20Draws%20Scrutiny.

57. Brandon Keim, "Uncle Sam Wants Your Brain," *Wired*, August 13, 2008, http://www.wired.com/wiredscience/2008/08/uncle-sam-wants/.

58. http://killology.com/Myth%20of%20returning%20vets%20&%20 violence.pdf.

59. Lawrence A. Farwell, "Brain Fingerprinting: A Comprehensive Tutorial Review of Detection of Concealed Information with Event-Related Brain Potentials," *Cognitive Neurodynamics* Volume 6, Number 2 (2012), 115–54, http://www.springer link.com/content/7710950336312146/.

60. Joseph A. Bernstein, "Seeing Crime Before it Happens," *Discovery*, January 23, 2012, http://discovermagazine.com/2011/dec/02-big-idea-seeing-crime-before-it-happens.

61. Sharon Weinberger, "Terrorist 'Pre-Crime' Detector Field Tested in United States," *Nature*, May 27, 2011, http://www.nature.com/news/2011/110527/full/news. 2011.323.html.

62. Canadians are probably most familiar with the case of Maher Arar, who has received both compensation for the Canadian government's alleged involvement in his detention, as well as apologies from both Canadian and U.S. governments.

63. Lawrence A. Farwell, "Brain Fingerprinting: A Comprehensive Tutorial Review of Detection of Concealed Information with Event-Related Brain Potentials," *Cognitive Neurodynamics* Volume 6, Number 2 (2012), 115–54, http://www.springer link.com/content/7710950336312146/.

64. The Human Genome Project was only to identify and sequence of all the genes in DNA, it does not address specifically what each gene does. Knowing where each gene does, however, help the myriad of projects working on to link specific genes with specific traits, functions, and conditions.

65. Federal Bureau of Investigation, "Amerithrax or Anthrax Investigation,", http://www.fbi.gov/about-us/history/famous-cases/anthrax-amerithrax/amerithrax-investigation.

66. Gregory D. Koblentz, *Living Weapons* (Ithaca: Cornell University Press, 2009), 212–14.

67. Ibid., 67–70.

68. Centers for Disease Control and Prevention, "Researchers Reconstruct 1918 Pandemic Influenza Virus; Effort Designed to Advance Preparedness," Press Release, October 5, 2005, http://www.cdc.gov/media/pressrel/r051005.htm.

69. Elizabeth Pennisi, "Synthetic Genome Brings New Life to Bacterium," *Science*, May 21, 2010, Vol. 328, no. 5981, 958–59, http://www.sciencemag.org/con tent/328/5981/958.full

70. "IGEM/Learn About," September 26, 2011, http://igem.org/wiki/index. php?title=IGEM/Learn_About&oldid=4750.

71. Although *E. coli* is responsible for several food scares, it also happens to be a useful platform for synthetic biology and genetic engineering experiments, including those conducted by hobbyists.

72. Rebecca Boyle, "How Modified Worms and Goats Can Mass-Produce Nature's Toughest Fiber," *Popular Science*, June 10, 2010, http://www.popsci.com/science/article/2010–10/fabrics-spider-silk-get-closer-reality.

73. U.S. Navy, "First Navy Trainer Completes Biofuel Flight at Patuxent River," August 25, 2011, http://www.navy.mil/search/display.asp?story_id=62384.

74. Phil McKenna, "A New Route to Cellulosic Biofuels," *Technology Review*, November 20, 2009.

75. Helen Knight, "E. coli Engineered to Make Convenient 'Drop-In' Biofuel," *New Scientist*, July 29, 2010, http://www.technologyreview.com/energy/23989/http://www.newscientist.com/article/dn19238-e-coli-engineered-to-make-convenient-dropin-biofuel.html.

76. Austin Wright, "Navy Powers Up Campaign for Great Green Fleet," *Politico*, March 7, 2012, http://www.politico.com/news/stories/0312/73752.html.

77. In the aftermath of the 9/11 attacks, some in the energy-conservation movement were making direct linkages between U.S. energy imports from the Middle East and terrorism.

78. Congressional Budget Office, *Energy Security in the United States*, May 9, 2012, http://www.cbo.gov/sites/default/files/cbofiles/attachments/05–09-EnergySecurity.pdf.

79. Ibid.

80. Noah Shachtman, "The Army's New Land Warrior Gear: Why Soldiers Don't Like It," *Popular Mechanics*, October 1, 2009, http://www.popularmechanics.com/technology/military/4215715.

81. Matthew Cox, "SF Units to Get Latest Land Warrior Kit," *Army Times*, November 9, 2009, http://www.armytimes.com/news/2009/11/army_landwarrior_110909w/.

82. DARPA, "Darpa Researchers Design Eye-Enhancing Virtual Reality Contact Lenses," January 31, 2012, http://www.darpa.mil/NewsEvents/Releases/2012/01/31.aspx.

83. Michio Kaku, *The Physics of the Future* (New York, Double Day, 2011), 42.

84. Clay Dillow, "DARPA Invests in Megapixel Augmented-Reality Contact Lenses," *Popular Science*, February 1, 2012, http://www.popsci.com/technology/article/2012–02/video-nano-enhanced-contact-lens-makes-augmented-reality-more-realistic.

85. Eye tracking is actually reaching application as it is one method used to sense how alert someone is.

86. Popular Science, *21st Century Soldier* (Time Inc., 2002), 122.

87. Eurofighter Consortium, "Direct Voice Input," http://www.eurofighter.com/media/press-office/facts-sheet-mediakit/direct-voice-input.html.

88. John Schutte, "Researchers Fine-Tune F-35 Pilot-Aircraft Speech System," October 15, 2007, http://www.af.mil/news/story.asp?id=123071861.

89. Michael P. McLoughlin, "DARPA Revolutionizing Prosthetics 2009," presented at the MORS Personnel and National Security Workshop, January 2009, http://www.dtic.mil/cgi-bin/GetTRDoc?Location=U2&doc=GetTRDoc.pdf&AD=ADA519193.

90. Sally Adee, "Winner: The Revolution Will Be Prosthetized," *IEEE Spectrum*, January 2009, http://spectrum.ieee.org/robotics/medical-robots/winner-the-revolution-will-be-prosthetized.

91. Andrew Zajac, "Robotic Arm Getting a Hand from the FDA," *Los Angeles Times*, February 9, 2011, http://articles.latimes.com/2011/feb/09/nation/la-na-prosthetic-arms-20110209.

92. Ibid.

93. U.S. Department of Energy Office of Science, "Artificial Retina Project," http://artificialretina.energy.gov/index.shtml.

94. General Electric, "The Story Behind the Real 'Iron Man' Suit," November 23, 2010, http://www.gereports.com/the-story-behind-the-real-iron-man-suit/.

95. Ibid.

96. Berkeley Robotics & Human Engineering Laboratory, "BLEEX," http://bleex.me.berkeley.edu/research/exoskeleton/bleex/.

97. Erico Guizzo and Harry Goldstein, "The Rise of the Body Bots," *IEEE Spectrum*, October 2005, http://spectrum.ieee.org/biomedical/bionics/the-rise-of-the-body-bots.

98. Ibid.

99. David Axe, "Combat Exoskeleton Marches toward Afghanistan Deployment," *Wired*, May 23, 2012, http://www.wired.com/dangerroom/2012/05/combat-exoskeleton-afghanistan/#more-81313.

100. B.S. Richardson, editor. *Phase I Report: Darpa Exoskeleton Program*, Oak Ridge National Laboratory, 2004, http://www.ornl.gov/~webworks/cppr/y2004/rpt/118274.pdf.

101. The Associated Press, "Army Bans Use of Privately Bought Armor," March 30, 2006, http://www.usatoday.com/news/washington/2006–03–30-bodyarmor_x.htm.

102. Daniel McIntosh, "The Transhuman Security Dilemma," *Journal of Evolution & Technology*, December 2010, Vol. 21 Issue 2, http://jetpress.org/v21/mcintosh.htm.`

Small Satellites

Related to ubiquitous space access is the subject of small satellites. Although trying to predict the future of space flight is often a futile act, if ubiquitous space launch is to be achieved, it will likely be initially capable of only launching small payloads. Advances in electronics have allowed more capability to fit into smaller satellites, making practical many missions that otherwise would be unaffordable. Today, the low cost of microsatellite construction has allowed a great proliferation in the number of nations (as well as universities) having their first satellites placed into orbit. These "first" satellites are making their way to orbit often by filling up unused capacity of other missions, piggybacking to orbit as possible within the constraints of the existing space-launch market.

Small satellites also have the potential to influence the shape of military matters in space. Similar to producing supercomputers from networked clusters of mass-produced computers (such as early model Playstation 3 gaming consoles[1]), for some missions a formation of small satellites can be combined to provide the services that otherwise would require a single satellite. Defense Advanced Research Projects Agency (DARPA) is pursuing this option under the Future, Fast, Flexible, Fractionated, Free-Flying Spacecraft United by Information Exchange (System F6) program.[2] This is a technology-development program. It is not meant to provide specific military capabilities, but instead to provide an architecture on which military capabilities, such as tactical surveillance, can be built. Among DARPA's goals for System F6 is to give each satellite a high degree of on-orbit maneuverability. Officially, it means the "capability to perform a semiautonomous defensive cluster scatter and regather maneuver to rapidly evade a debris-like threat."[3] Even without overt hostile action, the emergence of ubiquitous orbital access will increase congestion and the problem of space junk. The only difference between accidents and warfare is the lack of intent in the former.

Weaponized small satellites, due to their inherent low cost, are a prominent option that becomes all the more viable with ubiquitous space access. Small satellites as weapons are also a far-from-new idea. Before the small, micro, and nano-sat monikers became fashionable, small satellites and mass low-cost space launch made up an early deployment plan for space-based ballistic missile defense. During the later parts of the U.S. strategic defense initiative (SDI), more commonly known by detractors and some supporters as "Star Wars," the space-based kinetic-kill vehicle (SBKKV) concept of having several interceptor missiles onboard a few dozen large satellites, effectively orbiting missile batteries, changed into a plan to maintain in orbit thousands of small satellite-sized interceptors, known as *Brilliant Pebbles*. Each *Brilliant Pebble* was a self-contained kinetic energy (ramming) interceptor, launched in numbers to provide both global coverage, and an overwhelming number of targets for Soviet antisatellite (ASAT) capabilities to counter. The original Reagan-era *Brilliant Pebble* and later scaled back *Global Protections against Limited Strikes* (GPALS) program of the George H. W. Bush administration were a desired military capability that pointed toward low-cost space systems and access, and factored into some advance launch vehicle research of the day.[4] Similar to the collapse of the same era's portable satellite phone industry to ground-based cellular technology, space-based missile defense was supplanted by ground-based missile defense.

On a smaller scale than protecting the world from Soviet intercontinental ballistic missile (ICBM) attack, is the concept to use a satellite to protect another satellite, a concept referred to as a bodyguard satellite. Without the need for global coverage, a much smaller number of bodyguard satellites orbit in formation with the satellite being protected. On detection of a threat, the bodyguard satellite could impose itself between the ASAT and the satellite it is protecting. The basic concept does not address the matter of the debris such a "save" would produce. Also the term defensive satellite is contentious as it only reflects a matter of intent, the same small satellite could instead of protecting a satellite, be used to attack another satellite. Expanding the concept from protecting individual satellites to proactive control of what may reach orbital altitudes leads back to comprehensive global ballistic missile defense concepts.

Notes

1. Bryan Gardiner, "Astrophysicist Replaces Supercomputer with Eight PlayStation 3," *Wired*, October 17, 2007, http://www.wired.com/techbiz/it/news/2007/10/ps3_supercomputer.

2. DARPA, "System F6," http://www.darpa.mil/Our_Work/TTO/Programs/System_F6.aspx.

3. Ibid.

4. Andrew J. Butrica, *Single Stage to Orbit: Politics, Space Technology, and the Quest for Reusable Rocketry* (Baltimore, MD: The Johns Hopkins University Press, 2003), 117.

Military Space Planes

The X-37B Orbital Test Vehicle (OTV) program ends the United States Air Force's long wait to finally have its own fleet of space planes. By July 2012, each of the two 8.9-meter-long[1] unmanned space planes had each flown lengthy orbital missions and made autonomous landings. Although the flight to orbit via *Atlas V* rockets could not be hidden, once in orbit the missions of each OTV have been shrouded in secrecy, fuelling speculation. From the beginnings of the Space Age, there has been a line of thought that space is an extension of air power and many see this program as a potential means for the USAF to expand the military space to more broadly resemble the weaponized nature of air power. On the other hand, space is a different environment, and the winged X-37 vehicles may not necessarily be replicating traditional USAF warfighting roles in space.

Originally the X-37 was a NASA program to demonstrate technologies that may be useful for future spacecraft, with emphasis on recovery technologies.[2] In 2004, the program was transferred to Defense Advanced Research Projects Agency (DARPA), which the completed in-atmosphere testing of the design in 2006.[3] These earlier craft in the X-37 program had been air-dropped to demonstrate autonomous flight control of the unpowered vehicle. Like the much larger space shuttles that were in service at the time, the X-37 program always envisioned gliding to a runway after orbital reentry. This manner of recovery has specific costs and benefits. Once the airborne testing was completed the USAF proceeded with orbit-capable spacecraft in the form of the pair of OTVs.

For early manned spaceflight and cargo recovery (such as returning film from reconnaissance) satellites, ballistic capsule reentry proved simpler to launch, but required expensive infrastructure to pick up the capsule after reentry. During the early Space Age there were parallel efforts to construct aircraft capable of reaching space. The X-15 rocket plane program boosted several test pilots up to altitudes that qualified them for astronaut wings.[4] The air-launched X-15 was expected to be followed by the vertically launched X-20 Dynasoar program, which was conceived in the 1950s as a space bomber, and later became an orbit-capable reusable spacecraft useful for various military missions, in addition to the delivery of nuclear warheads. X-20 was cancelled in 1963. Prior to X-20 there were the similar BoMI (bomber missile) and Project ROBO (rocket bomber) that also investigated a U.S. rocket-powered boost-glide weapon system. Inspiring all this was Eugene Sänger's "Silverbird" concept for a suborbital "Amerika Bomber" from World War II. Another Nazi Wunderwaffe concept, the "Silverbird" was to be a horizontally launched boost-glide bomber to fulfill Hitler's desire to strike the United States. Elements of the Soviet leadership were also convinced that the U.S. Space Shuttle, which at the time was a dual NASA and DOD program, was actually a space bomber.[5] It is no surprise that among the speculated roles for the pair of USAF spacecraft is attack. That said, the X-37B concept, despite its ability to maneuver would seem to be a very expensive strike option, even taking into account the experimental nature of Prompt Global Strike (PGS) work thus far. Any weapon deployed by an orbiting spacecraft, such as X-37B, would have to either survive reentry at orbital speeds, or carry propulsion to slow down—it must be noted that the size of an X-37B payload bay is about that of a pickup truck.[6]

Thus far, flown PGS-related hardware have all been suborbital hypersonic glide vehicles with some degree of flight control, essentially advance forms of the shapes used in existing nuclear weapon reentry vehicles. This in itself has caused some concern in that similarities between PGS and existing U.S. nuclear strike systems would lead to these two different weapon systems being confused for one another by a third-party nuclear-armed nation, causing a mistaken retaliatory strike, by actual nuclear weapons, on the United States.[7] The emphasis is to extend the glide portion of a PGS weapon's flight, as currently deployed U.S. nuclear warheads lack such capabilities.[8] That said, for all the complaints from near peers, and others, about potential for confusion between PGS and a nuclear strike, other nations are deploying maneuverable reentry vehicle (MaRV) warheads: Russian claims that their upgraded nuclear deterrent is able to evade U.S. missile defense, and China's antiship ballistic missile (ASBM) both point toward the use of a MaRV, especially the ASBM as it would need to seek after a moving target after its several-minutes-long flight.

Proponents of air-breathing space-launch propulsion have their own solution to concerns that a rocket-boosted PGS would be confused with a nuclear strike—an actual hypersonic bomber. Eugene Sänger's "Silverbird" concept was a track-launched bomber that would be initially propelled by rocket and proceed across the globe from Germany to Japan by skipping across the atmosphere. As a prelude to today's rocket versus space plane competition, there was a competing Nazi scheme to attack the U.S. homeland with ballistic missiles—the von Braun's A9/A10 multistage missiles. Both Sänger and von Braun fell into Allied hands after the war, with the latter being more successful so far at least. Hypersonic aircraft development in the United States has been ongoing since the Cold War, with the X-30 National Aero-Space Plane (NASP) being the best known in the public domain. Recent USAF planning documents do make a link between air-breathing launch vehicle technology and a hypersonic bomber, though as a means of leveraging off of each other's development as a cost-saving measure for what can only be two very expensive endeavours.[9] Although orbit-to-surface strike cannot be ruled out as an X-37B mission, it does seem unlikely due to the costs versus benefits.

The other offensive mission speculated on for the pair of OTVs is satellite inspection—often used as a euphemism for antisatellite attack. Satellite inspection and some ASAT attack profiles require on-orbit rendezvous, something that it should be noted is not listed as among the X-37's capabilities, though as mentioned earlier, there has not been much released as to specific on-orbit activities of either missions flown as of July 2012. Known autonomous on-orbit rendezvous experiments, such as the XSS-11, have generated quite a bit of controversy. Basically any time a spacecraft can operate safely in close proximity with another spacecraft there is potential for it to conduct inspection and attack.

Another potential role is for the X-37Bs to act as a satellite bus for reconnaissance and surveillance payloads. These spacecraft are able to maneuver on-orbit across missions lasting hundreds of days, meaning that they have the opportunity to surprise those on the ground through unexpected changes regarding the period during which it will be overhead. Unlike a conventional spy satellite, the winged reentry capability allows the X-37 to return the entire payload to earth for later reuse. As each vehicle has only flown once, not much is known about turnaround times, though the fact that they are launched by conventional rockets means that they are not operationally reactive in the sense that they can launch on demand. Once in orbit, however, the capacity to change orbits does give it some operational flexibility.

Finally there is the possibility that the U.S. government is being completely truthful, though vague, and the X-37B program is purely to test new space and orbital reentry technologies. It must be remembered that aside from the Soviet Buran no space plane has returned to a runway landing unmanned. By July 2012, this feat had been accomplished twice by the USAF. By definition the International Space Station (ISS) would be an inappropriate venue for DOD-specific experiments in orbit, and without the space shuttle, recoverable long-duration military experiments would have few opportunities to fly. Indeed with an X designation, the 224 and 469 day[10] missions of the first and second OTVs would seem to be endurance experiments in themselves. Now of course experience gained during the X-37B program may be applicable toward other emerging military space capabilities.

Notes

1. United States Air Force, "Factsheets: X-37B Orbital Test Vehicle," March 3, 2011, http://www.af.mil/information/factsheets/factsheet.asp?fsID=16639.

2. National Aeronautics and Space Administration. "NASA—X-37 Fact Sheet (05/03)," June 2003, http://www.nasa.gov/centers/marshall/news/background/facts/x37facts2.html.

3. Ibid.

4. National Aeronautics and Space Administration. "X-15—Hypersonic Research at the Edge of Space," February 24, 2000, http://history.nasa.gov/x15/cover.html.

5. Bart Hendrickx and Bert Vis, *Energiya-Buran: The Soviet Space Shuttle*, (Chichester, UK: Springer, 2007), 53–55.

6. Stephen Clark, "Air Force Spaceplane Is an Odd Bird with a Twisted Past," April 2, 2010, http://spaceflightnow.com/atlas/av012/100402x37update/.

7. David Axe, "New 'ICBMs vs. Terrorists' Plan: Now 50% Less Crazy!" *Wired*, March 7, 2011, http://www.wired.com/dangerroom/2011/03/new-icbms-vs-terrorists-plan-now-50-less-crazy/.

8. Ibid.

9. United States Air Force. *Report on Technology Horizons: A Vision for Air Force Science & Technology During 2010–2030, Volume 1*, May 15, 2010, http://www.af.mil/shared/media/document/AFD-100727–053.pdf.

10. United States Air Force, 30th Space Wing Public Affairs, "X-37B Orbital Test Vehicle lands at Vandenberg," June 16, 2012, http://www.vandenberg.af.mil/news/story.asp?id=123306242.

Competition to Directed Energy Weapons

If directed energy weapons (DEWs) represent a Revolution in Military Affairs (RMA) in waiting, then there must be a traditional technological and the doctrinal paradigm struggling to remain current. With a constrained funding environment, the biggest challenge to the operational deployment of DEW systems may come from simple refinements to existing weapons' concepts. A major variable in this competition between bullets, bombs, and missiles versus energy beams is the future of electronics, and micro-electro-mechanical systems (MEMS). Electronics have so far shown a tendency to lower costs, while becoming more capable, and have already been applied to the existing world of physically delivered destruction.

Electronically guided and triggered munitions are getting smaller. There are, for instance, several programmable fuse airburst 40-mm grenade systems on the market meant to explode when over or next to a target—defeating cover. The U.S. XM25 magazine-fed grenade launcher has taken this concept down to a 25-mm grenade.[1] Internationally, South Korea is deploying the K11, a magazine-fed 20-mm airburst grenade launcher integrated with an assault rifle.[2] Now these are still ballistic systems in that the electronics are only there to set off the grenade; the shooter still has to aim the weapon to get the grenade near enough for the airburst to be effective. On a larger scale, electronic fusing is allowing for controlled detonation of larger warheads, able to shape warhead effects—enhance blast and fragmentation or to produce an armor-penetrating stream of metal.

Precision control over how energetic materials burn or explode may give the gun all the precision desired out of DEW systems. For instance, advances in controlling muzzle velocity could make guns more usable as less-than-lethal weapon. Blunt force trauma can be lethal if used incorrectly as can be seen in numerous incidents involving baton and beanbag rounds. Safe operation of these weapons requires knowing when the target is too close. Alternatively, a reliable means to gauge distance, and bleed off the right amount of propelling gases would allow fine control over the impact velocity of less-than-lethal rounds. The principle of controlling velocity by regulating chamber pressure is already in use in paintball markers. As of early 2010, Lund Technologies has been demonstrating this concept in their *Lund Variable Velocity Weapons System* (LVVWS) with the help of U.S. small business innovation research funding.[3]

Somewhat further off in the development pipeline is the "smart bullet" concept, where a round's trajectory could be affected by onboard means, MEMS for aerodynamic steering, or control of small rocket propulsion. For all intents and purposes this sci-fi concept is based around the premise that shrinking electronics will eventually allow a missile to be bullet size. Although not offering the same "novel effects," guided small-caliber cannon shells and perhaps even bullets offer a degree of precision comparable to that of DEWs. Leading toward this are the many defense companies offering laser-guidance kits for 2.75-inch (70 mm) diameter rockets such as the U.S. Hydra 70, or more commonly in Western militaries outside the United States, the Canadian-made CRV7. In this sense, the ongoing implosion of electronics would make DEWs an unneeded revolution.

Notes

1. XM25 is an outgrowth of the less-successful XM29 Objective Individual Combat Weapon program, which combined a 20-mm-diameter grenade system with a select fire rifle.

2. The K11 resembles the cancelled XM29, though the grenade launcher is bolt action instead of autoloading. This, however, does reduce the overall weight of the weapon.

3. Katie Drummond, "Inventors Design Less-Lethal 'Taser Me Elmo' Rifle," *Wired*, March 22, 2010, http://www.wired.com/dangerroom/2010/03/toy-firm-designs-less-lethal-taser-me-elmo-rifle/.

Electromagnetic Railguns

The electromagnetic (EM) railgun in many respects sits in the fuzzy boundary between conventional and emerging military technologies. It is an emerging technology in that it is yet not practical, but does promises much. However, in many ways it is offering more of the same. Compared to conventional bullets, shells, and most tactical missiles, it offers a revolution in speed, indeed the point of much of this work is to push a projectile to speeds where the destructive power of the projectile's kinetic energy (KE), its energy of motion, is greater than that which can be carried in the projectile. On the other hand these supersonic and hypersonic velocities are not comparable to the light-speed performance of EM weapons such as lasers. Like directed energy weapon (DEW) systems, it faces competition from existing technologies that can achieve many of the same effects, but without the many technical challenges that may not be overcome in the near and medium term.

In conventional guns, the bullet or shell is propelled by expanding gases from the rapid combustion of a propellant such as gunpowder, and are limited by the physics of the burning propellant and the resulting gas. Among the limitations is that the propellant burn in a gun is deflagration, which is burning at or below the speed of sound within the material. A detonation, combustion propagated by a shockwave that is above the speed of sound in the material, would be counterproductive as it would turn the gun into an expensive pipe bomb. Chemical-powered guns can overcome these limitations by using the deflagration process to drive a piston that compresses a large quantity of light gas, such as helium, which when released can achieve higher velocity than the combustion gases can achieve. Light gas guns are found in laboratories that study high-velocity impacts, such as in meteorite impact research, but are impractically large and cumbersome to be useful as weapons.

In an EM railgun, a powerful electric current is run from one rail to another through a conductive portion of the projectile, the armature, generating a powerful Lorentz force, which accelerates the projectile along the two rails. Not being limited by the characteristics of a chemical propellant, EM railguns have achieved hypersonic (above Mach 5) muzzle velocities from relatively small guns.[1] The KE of a mass increases by the square of its velocity. Doubling the mass of the projectile merely doubles the KE. Doubling its velocity quadruples the KE. Therefore if one wants to use KE purely as a destructive mechanism, velocity at impact is paramount.

$$KE = 1/2 \ (Mass) \ (Velocity \ ^\wedge \ 2)$$

The extreme "muzzle" velocity of railguns has drawn military interest for years. Early in the Reagan-era Strategic Defense Initiative (SDI), orbiting railgun batteries were a possible means of launching guided KE interceptors. Earth-based applications, such as long-range artillery aboard warships, or even mobile ground pieces, avoid many of the power, mass, and volume restrictions found in space basing. As of 2012, working test beds, not laboratory experiments, were being demonstrated by BAE Systems and General Atomics[2] under the Office of Naval Research's Electromagnetic Railgun Innovative Naval Prototype (INP) program that started in 2005.[3]

Like DEW, cited applications for the U.S. Navy's railgun research include defense against missiles and aircraft; however, but unlike DEW, there is also open discussion about its use as a strike weapon. Unlike the novel effects of a laser or high-power radio-frequency weapon, the effects of being hit by a hypersonic projectile are unambiguously destructive. This of course puts this technology in competition with existing missile systems. Notwithstanding the research and development costs, the per-shot cost of a railgun would only involve a projectile and the energy source. Unlike electrically powered DEW systems, there would not be a "limitless magazine," a fact that puts the railgun conceptually in between existing gun and potentially revolutionary DEW systems.

Although the U.S. Army, U.S. Navy, and their contractors have produced spectacular media presentations hinting at the promise of this technology, it must be remembered that challenges remain. The weapons being worked on now are power hungry, requiring pulses of energy measured in multidozen megajoules. Naval power plants and ground-based generators are capable of recharging such system, but the sudden release of this type of energy requires advance capacitors, and other novel energy storage systems. The heat generated by this much energy being driven through a weapon system is destructive, and current goals involve developing cooling systems to allow for firing the railgun several times a minute. Then there is the problem of the rails themselves, they must be incredibly strong to survive the effects of firing. An armature must be in electrical contact with the rails, whether the armature remains solid or is intentionally transformed into superheated conductive plasma,[4] putting significant wear and tear on the rails. Present critics of this research have cited the short life span of current barrels, which have life spans measured in only hundreds of shots.[5] Finally there is the problem of constructing a guided projectile that can survive being fired from a railgun. Accelerating to hypersonic velocities in only meters creates high g-loads that would crush most available components today. Again there are the EM and thermal effects of EM railgun operation that would fry (in both senses of the word) electronics unless they were shielded.

Solutions to these challenges overlap with other emerging technologies. Electrically powered DEWs also require large pulses of energy to operate. Solutions to the energy problem may involve nanotechnology structures that can increase the energy density of capacitors and other forms of electrical energy storage. Barrel sustainability is in part a material-engineering problem, and again nanotechnology is a potential solution. Robust electronics that can survive the accelerations and hostile thermal and EM environment of being fired likewise combine advance electronics with nanotechnology.

That there are potential solutions to these challenges makes the EM railgun an emerging military technology instead of science fantasy. If they can be overcome, this technology's backers claim a potential military revolution of even lower cost precision firepower at longer standoff ranges. However, if railgun technology works as advertises it will still need to prove its advantages against existing weapons' concepts, such as the old-fashioned chemically driven guns and missiles. The excitement around emerging technologies is part about dreams of the future, and part about the thrill of the gamble—with the ironic payoff of becoming commonplace.

Notes

1. Grace Jean, Office of Naval Research. "With a Bang, Navy Begins Tests on Electromagnetic Railgun Prototype Launcher," February 28, 2012, http://www.onr.navy.mil/Media-Center/Press-Releases/2012/Electromagnetic-Railgun-BAE-Prototype-Launcher.aspx.

2. Ibid.

3. United States Navy, Office of Naval Research. "Electromagnetic Railgun," http://www.onr.navy.mil/en/Science-Technology/Departments/Code-35/All-Programs/air-warfare-352/Electromagnetic-Railgun.aspx.

4. David Hobbs, *An Illustrated Guide to Space Warfare* (New York: Prentice Hall, 1986), 121.

5. Spencer Ackerman, "Navy's Rail Gun Blasts through Budget Restrictions," *Wired*, February 10, 2012, http://www.wired.com/dangerroom/2012/02/rail-gun/.

Glossary

Active Denial System (ADS)—A less-than-lethal directed energy weapon that uses millimeter-wave radio-frequency technology to generate an "intolerable heating sensation."

Advanced Extremely High Frequency (AEHF)/MILSTAR 3—U.S. military communication satellite program to provide increased bandwidth, more security, and survivability for forces.

Advanced Tactical Laser (ATL)—U.S. research program to mount a kilowatt-class laser on a C-130 Hercules tactical transport aircraft for use against ground targets.

Airborne Laser (ABL)—A now-defunct U.S. program investigating the feasibility of using an aircraft-mounted megawatt-class high-energy laser for missile defense. The YAL-1 airborne laser test bed after a test program that included successfully intercepting in-flight liquid and solid-fuelled ballistic missiles was put into storage in early 2012.

Antiship Ballistic Missile (ASBM)—A long-range sea denial weapon being developed in the People's Republic of China. It is thought to be based around the DongFeng-21 (DF-21) medium-range ballistic missile with a maneuverable reentry vehicle able to attack moving warships.

Asimov's Three Laws of Robotics—Influential set of rules proposed by Isaac Asimov in several works of science fiction:

1. A robot may not injure a human being or, through inaction, allow a human being to come to harm.
2. A robot must obey the orders given to it by human beings, except where such orders would conflict with the first law.
3. A robot must protect its own existence as long as such protection does not conflict with the first or second laws.

These three laws have changed over time, and are sometimes joined by a zeroth law:

0. A robot may not harm humanity, or, by inaction, allow humanity to come to harm.

Ballistic Missile Intercept, Boost Phase—The ideal missile defense intercept, which targets a missile during its initial launch phase when it is most vulnerable (a single large, slow target that is easily detectable by its heat signature) as it rises from the ground prior to exiting the atmosphere and detaching from its boosters.

Ballistic Missile Intercept, Direct Ascent—Weapon flight profile where interception occurs at or near the apogee of a suborbital launch. Analogous to the flight of a research sounding rocket where the payload conducts it mission at the apogee of a suborbital launch. These are used by U.S. SM-3 and ground-based interceptor (GBI) ballistic missile defense systems.

Ballistic Missile Intercept, Midcourse—Interception of a ballistic missile after its engines have finished firing and the missile payload (warhead) is coasting though to its apogee and onward to the reentry, or terminal phase, of ballistic missile flight.

Biological Weapons Convention (BWC), 1972—Formally known as the Convention on the Prohibition of the Development, Production and Stockpiling of Bacteriological (Biological) and Toxin Weapons and on their Destruction. It is a long-standing multilateral treaty that the United States is party to; it prohibits the development, production, and stockpiling of biological and toxin weapons. It is also notable for its lack of verification mechanism.

Biometrics—The use of a person's unique physiological and behavioral traits and characteristics for identification.

Biomimetics—The application and adaptation of processes and functions found in biological systems in technological items. Examples include the replication of gecko surface adhesion via nanotechnology and the use of insect studies to produce flight systems suitable for use in micro unmanned air vehicles/unmanned air systems (UAVs/UASs).

Brain-Computer Interfaces (BCI)—The use of voluntary (controllable) brain activity as a means to control equipment.

Chemical Weapons Convention (CWC), 1993—Formally the Convention on the Prohibition of the Development, Production, Stockpiling and Use of Chemical Weapons and on their Destruction. It is a multilateral treaty that the United States is party to; it prohibits the development, production, and stockpiling of chemical weapons, as well as a commitment to destroy these agents, precursor chemicals, weapon systems, and related equipment.

Commercial Off the Shelf (COTS)—Government procurement practice of buying existing, in production, commercial goods instead of developing a government-specific version. The intention is to cut down on development costs, especially with software and computer hardware, if an acceptable commercial product is able fill a role.

Defense Advanced Research Projects Agency (DARPA)—Department of Defense agency noted for funding many seemingly outlandish concepts and projects. Conversely, the radical nature of DARPA programs often has its successes as world-changing technologies. An example of such would be DARPA's association with the early Internet.

Directed Energy Weapon—A means of attack where destructive energy is not transmitted to a target by physical means (ramming, fragmentation, or concussion). Generally refers to laser-, radio-frequency-, particle-beam-, and electromagnetic-pulse-based weapons, but may also include acoustic (sound) based weapons.

Electromagnetic Pulse (EMP)—A large burst of electromagnetic energy, often associated with nuclear explosions, but may be generated by nonnuclear and nonexplosive means.

Exoskeleton—In general, it is a hard external skeleton or shell that supports and protects an organism. In the context of this book, refers to powered, load-bearing mechanism that is worn to enhance carrying ability and strength.

Force Application and Launch from Continental United States (FALCON)—DARPA hypersonic research program. Under FALCON are programs to develop a small launch vehicle to support research and programs to construct and fly both rocket-launched and eventually runway-takeoff-and-landing hypersonic test vehicles.

Force Enhancement/Multiplication—Factors that greatly enhances a military's abilities versus an equally sized unenhanced force. These include technology, such as space assets and stealth technology, but may also include factors such as doctrinal changes, morale, and policy.

Force Enhancement, Space—The use of space-based assets to greatly enhance terrestrial warfare abilities in comparison to an unenhanced force.

Fractional Orbit Bombardment System (FOBS)—Soviet-era strategic strike system that put a warhead bus into a very low orbit with the intention that the warhead reentered the atmosphere before the completion o one orbit. FOBS's orbit and deorbit characteristics gives it ideal first-strike or surprise-attack capabilities. The status of such systems under the Outer Space Treaty (OST) ban on orbiting nuclear weapons remains to this day a matter of academic debate. Later arms-reduction treaties prohibited systems operating under the FOBS concept.

Fuel Air Explosives—Explosives that exploit atmospheric oxygen for part of its oxidizer, and therefore are dependent on dispersing the fuel component properly to achieve the correct fuel/air ratio before detonation. Also called thermobaric weapons, these weapons are noted for producing a large long-duration pressure wave.

High-Energy Laser (HEL)—In the military context a laser capable of producing destructive effects over significant ranges. Generally has power outputs in the hundreds of kilowatts to megawatts to achieve weaponized effects.

High-Power Microwave (HPM)—in the military context, a radio-frequency (microwave) weapon. With higher power outputs from some transmitters used by radars and radio-frequency countermeasures (radar and radio communication jammers), these devices often blur with dedicated high-power EM weapons.

Hybrid Rocket—A rocket engine where one of the propellants is solid and the other is either a gas or a liquid. Typically, it is the fuel that is a solid with a liquid or gas oxidizer. At present, these types of rockets offer a trade-off in performance in favor of low cost and safety.

International Traffic in Arms Regulations (ITAR)—U.S. regulations controlling the import and export of defense- and security-sensitive technologies.

Interoperability—It is the ability of different military agencies and militaries from different countries to work together cohesively. It requires not just hardware but also compatibility in force structures, doctrine, and training among other factors.

Kinetic Energy Weapon (KEW)—A weapon that employs the energy of an object in motion through collision to destroy another object.

Laser—Light amplification by stimulated emission of radiation (LASER), is both an effect predicted by quantum physics, and devices that make use of this effect to produce light that is of very directional nature and very coherent. Often lasers are described by their lasing medium, the medium that is stimulated into producing the laser light.

Laser, Chemical—A chemical reaction is used to power the excitement of the lasing medium. Many recent laser test beds have used chemically powered lasers due to their ability to generate high energy levels.

Laser, Free Electron—In a free electron laser, instead of atoms or molecules of a gas, liquid, or solid being excited, a beam of electrons is the lasing medium.

Laser, Gas Dynamic—Chemically powered laser that excites the lasing medium, a gas, via its expansion through nozzles after a supersonic (or near supersonic) flow. Not to be confused with gas lasers, which is a broader category that includes gas dynamic lasers, where the lasing medium is a gas.

Laser, Solid State—Generally electrically powered, lasers using a solid lasing medium such as neodymium doped glass. The first laser was a crystal of ruby excited by a flash lamp.

Laser, X-ray (Nuclear Bomb Pumped)—A weapons concept worked on under Strategic Defense Initiative (SDI). Reportedly, it was to project multiple high-energy laser beams in multiple directions, all powered by the detonation of a nuclear device.

Basing concepts included ground- and sea-based direct-ascent boosters, and on-orbit mines (though the use of a nuclear detonation as the power supply led to arguments against this basing concept due to possible conflict with the 1967 OST).

Launch Vehicle—A vehicle (rocket, missile) designed to move a payload into or through space. With respect to orbital space launch, it must be able to overcome gravity and accelerate a payload typically from a standstill to an initial orbital velocity needed for stable orbit.

Launch Vehicle, Expendable—A launch vehicle that is used once with no effort made to recover it for further use.

Launch Vehicle, Hybrid—A multistage, partially reusable, launch vehicle in which the initial stage is reusable and later stages are expendable. The first stage does not reach orbital velocities and therefore has greatly reduced costs (both financial and technical) associated with recovery. This term is also used to describe a launch vehicle that uses hybrid rocket propulsion.

Launch Vehicle, Reusable—A launch vehicle that is recovered for reuse. Scaled composites' fully reusable *Space Ship One/Tier One* is a suborbital vehicle only. The U.S. Space Shuttle, Soviet Buran/Energia system, and SpaceX's *Falcon 1* are the only partially reusable orbital launchers to have attempted spaceflight. At present, there are no examples of fully reusable orbital launch vehicles.

Less-Than-Lethal Weapons—These are weapons meant to minimize lethal effects, often used as a manner to achieve compliance without having to use lethal force. The term nonlethal is also used; however, this term may give the impression that these weapons have zero lethal potential. Accidental or incorrect use of less-than-lethal systems may cause serious injury and even death.

Mach Number—Ratio between the speed of an object and sound under the particular environmental conditions present for the object. Space does not present an adequate medium for sound transmission; hence there cannot be a speed of sound in space. Spacecraft velocities are often made in comparison to terrestrial Mach numbers (at sea level, 1,225 kilometers per hour), with Mach 25 being the necessary speed to reach a minimum orbit.

Mid-Infrared Advanced Chemical Laser (MIRACL)—U.S. experimental laser facility located at the White Sands Missile Range, New Mexico. This is a megawatt-class chemical laser, which in the course of its use in experiments has demonstrated latent antisatellite (ASAT) capabilities. In 1997, with much controversy over its weaponization potential, MIRACL was used to test the vulnerability of U.S. satellites to laser attack by briefly illuminating a soon to be out of service U.S. Air Force satellite.

MILSTAR—U.S. military communication satellite program. The AEHF (MILSTAR 3) program is the third generation of satellites to carry the MILSTAR name.

National Aeronautics and Space Administration (NASA)—U.S. space agency responsible for civilian space exploration, among other missions or purposes.

NAVSTAR GPS—U.S. Global Positioning System. NAVSTAR GPS provides not only free accurate positional data, but also an accurate timing signal on which most electronic financial transactions are synchronized. Accurate location and accurate timing data also imply accurate motion data for moving objects equipped to receive GPS, allowing for its use in precision-guided munitions (PGMs). It emits two signals; the less accurate public C/A-code for general use and the encrypted P-code for military use.

Operational Responsive Space (ORS)—A movement to reduce the cost and time needed to identify and deploy a new space capability, usually in the context of space-force enhancement. Technologies associated with ORS include microsatellites; low-cost, small-payload launch vehicles; and near-space aerial vehicles (as low-cost substitute for orbiting satellites).

Orbit—To completely circle a body. The unpowered or free-flight state where bearing other forces a spacecraft will circle the earth indefinitely due to a combination of gravity and momentum imparted earlier by rockets.

Parasite Spacecraft (Space Mine)—A satellite sent to orbit in formation with an opposing power's satellite, ready to conduct some action on command. Often in the context of ASAT spacecraft but can also be used in the form of less-active means, such as surveillance.

Precision-Guided Munitions (PGMs)—Munitions with onboard terminal guidance capable of hitting exceedingly close to the aim point. Also referred to as smart or brilliant weapons.

Prompt Global Strike—The capability to strike any point on earth with nonnuclear weapons in time measured in hours and minutes rather than days. Candidate technologies for this as of yet unfulfilled requirement include hypersonic transatmospheric vehicles and maneuvering PGMs delivered by suborbital reentry vehicles.

Responsive Space Launch—The capability to launch small payloads into orbit on demand. Related technologies include low-cost space access, space tourism, operationally responsive space, hybrid launch vehicles, air-launched launch vehicles, single-stage-to-orbit launch vehicles, and streamlining overall space access.

Revolution in Military Affairs (RMA)—Often used to describe paradigm shifts in military power whether brought on by operational or technological changes. In the present context, refers to the Information Age doctrines and hardware, which it is argued has brought about a new way to fight wars.

Rocket—A self-contained propulsion system that works by expelling propellant in a direction opposite to the direction of flight. At its most basic, a rocket is an implementation of Newton's third law of motion: for every action, there is an equal and opposite reaction. A rocket ejects via some kind of energy source a reaction mass in one direction, resulting in thrust in the opposite direction. In a chemical rocket, the propellants (oxidizer and fuel) both generate the energy and provide the reaction material. Thermal rockets have an external heat source (nuclear, solar, beamed energy, etc.) that provides energy to expel reaction mass. Electric rockets use electrical and/or magnetic principles to accelerate reaction mass, often in the form of charged particles.

Rocket, Artillery, and Mortars (RAM)—A term used to collect low-cost ballistic threats that have been employed by subnational actors such as terrorist groups against military and civilian targets. Responses to the RAM threat have included sending in military forces to occupy areas from which they are launched, and the development of defensive weapons, such as interceptor missiles and lasers, that are able to respond to salvos of these weapons within their short flight time.

Shutter Control—In the context of this discussion, regulatory and legal means imposed on companies offering remote imaging services to control the flow of such data to potential hostile powers.

Signals Intelligence (SIGINT)—In the context of this discussion eavesdropping on electronic emission on the earth's surface via satellites, and on communications between satellites and ground stations.

Small Satellites—As computers become increasingly powerful and smaller, more capability can be fitted into smaller payloads. One result has been the development of smaller satellites, which may be further classified as follows:

- Minisatellite, 100–500 kilogram
- Microsatellite, 10–100 kilogram
- Nanosatellite, less than 10 kilogram
- Picosatellite, less than 1 kilogram

There is no official or standardized terminology for classifying satellites by size.

Space-Based Laser (SBL)—An orbiting antiballistic missile, high-energy laser weapon concept that is regularly proposed but as of yet unfunded. A constellation of roughly two dozen orbiting laser weapon satellites could give constant coverage to a wide band of the earth.

Space Militarization—The use of space assets for military purposes. These range from Cold War–era strategic surveillance to present-day force enhancement and, in the future, force application. It is currently distinguished from the weaponization of space, which by legal default consists of deploying a conventional weapon on orbit.

Space Sanctuary—Strategy where space is kept weapons free (ASAT free) through norms and agreements, so as to allow its use for strategic surveillance, early warning (in the context of nuclear war), and space-force enhancement.

Space Weaponization—A rather contentious term, in general it refers to a situation where space-force application is widely practiced by one or more nations. At present due to latent space weapons capabilities found in nuclear-armed missiles (fused to go off at high altitude), space weapons experiments (such as MIRACL laser facility at White Sands, New Mexico), and the fact that ballistic missiles do transit through space, the distinction between today's state of space militarization and weaponization can be described as fuzzy.

Staging, Multi—To improve the payload and structure versus fuel mass ratio. Staging is used to shed dead weight (by dropping spent stages) as a launch vehicle ascends. Tsiolkovski's rocket equation provides the mathematical rationale for multistage rockets.

Staging, Single—While staging offers relaxed mass fractions for payload, it is operationally expensive due to the complexity of having essentially multiple vehicles that all must operate correctly. For a rocket-powered single stage to orbit vehicle, it is not uncommon to find the propellant between 80 and 90 percent or more of the total launch mass. Air-breathing propulsion allows for lower propellant requirements as the oxidizer is largely supplied by the environment.

Starfish Prime—This was the U.S. test in 1962 of a 1.4-megaton warhead at an altitude of 248 miles over the Pacific Ocean, causing a blackout of communications over the area and permanently damaging three satellites in orbit. In 1963, such tests would be prohibited with the Limited Test Ban Treaty.

Suborbital—A spacecraft that has at least achieved the nominal definition of space (the Kármán line at 100-kilometer altitude) but does not have the momentum needed to complete one orbit and is pulled back down by gravity.

Suborbital Space (sometimes referred to as near-space)—Region between the limits of aerodynamic flight (approximately 50-kilometer altitude) and the minimum altitude where an unpowered orbit will not rapidly decay (approximately 150 kilometers). This region is often referred to as a no-man's land between aeronautics and spaceflight, as it very difficult to do more than simply traverse this area on rocket thrust. Very-high-altitude ballooning (in the lower parts of this region) and some transatmospheric vehicle technology (for the higher end) may allow exploitation of this domain.

Supervisory Control and Data Acquisition (SCADA)—Computer software and hardware that form an industrial control system. Presently of interest to security and defense circles as the use of common computer hardware and software, including remote control applications such as remote desktop software, has opened potential vulnerabilities to cyber attack on modern industry and critical infrastructure such as power and water supply.

Tactical High-Energy Laser—Joint U.S. and Israeli program to develop a laser-based defense system capable of defeating RAM and other shorter-range threats. It has successfully destroyed both in-flight rockets and mortars, including salvos. Operational difficulties have lead to this program being cancelled after a series of high-profile tests.

Unmanned Aerial Vehicle (UAV)—In contemporary terms refers to semiautonomous aircraft. For beyond line-of-sight operations from ground controllers, UAVs depend on satellite communications to relay back data and to receive commands. GPS navigation is also important for UAV operations.

Unmanned Air System (UAS)—A newer term that includes both the UAV, as well as its ground station. Examples include hand-launched micro UAVs and the laptop or smaller computer device used to control it.

Unmanned Combat Aerial Vehicle (UCAV)—Armed UAV aircraft. While ad hoc antiarmor-armed predator UAVs may be considered UCAVs, this term generally is used to describe a dedicated autonomous warplane able to reach and engage targets with little human supervision. Related to it is the unmanned combat air system, which like the UAV/UAS relationship includes the ground infrastructure needed to support the UCAV.

X-20 Dynasoar—U.S. manned space plane project that ran from 1957 to 1960. It was to be vertically launched by expendable multistate rockets and landed horizontally for reuse.

X-37—U.S. autonomous space plane project that is ongoing. Formerly a NASA program, as a USAF program, two X-37B orbital test vehicles have been procured and launched on missions lasting months each. It is vertically launched by expendable *Atlas* V launch vehicles and lands horizontally for reuse.

X-41 Common Aero Vehicle (CAV)—U.S. reentry vehicle project meant to allow for a wide degree of cross-range maneuvering. Reportedly it is meant for a suborbital launch and is under consideration for a U.S. prompt global strike capability.

X-47 Unmanned Combat Air System Demonstrator (UCAS-D)—U.S. Navy program to develop an UCAS able to conduct operations off the deck of an aircraft carrier. Additional goals include an autonomous aircraft able to refuel in flight to increase this weapon system's range and endurance.

Bibliography

Primary

Alston, Philip. United Nations Human Rights Council. *Report of the Special Rapporteur on Extrajudicial, Summary or Arbitrary Executions, Philip Alston,* May 28, 2010. http://www2.ohchr.org/english/bodies/hrcouncil/docs/14session/A.HRC.14.24.Add6.pdf.

Barr, Larine. 88th Air Base Wing Public Affairs, United States Air Force. "Pulsed Detonation Engine Flies into History." *Air Force Print News Today,* May 16, 2008. http://www.afmc.af.mil/news/story_print.asp?id=123098900.

Beard, Jonathan. "DARPA's Bio-Revolution," *DARPA: 50 Years of Bridging the Gap.* http://www.darpa.mil/About/History/First_50_Years.aspx.

Brandler, Philip. "The United States Army Future Force Warrior—An Integrated Human Centric System." http://ftp.rta.nato.int/public//PubFullText/RTO/MP/RTO-MP-HFM-124///$MP-HFM-124-KN.pdf.

Caldwell, John A. "Dextroamphetamine and Modafinil are Effective Countermeasures for Fatigue in the Operational Environment." http://ftp.rta.nato.int/public//PubFullText/RTO/MP/RTO-MP-HFM-124///MP-HFM-124-31.pdf.

Callier, Maria. Air Force Office of Scientific Research. "AFOSR and NASA Launch First-Ever Test Rocket Fueled by Environmentally-Friendly, Safe Aluminum-Ice Propellant," August 21, 2009. http://www.wpafb.af.mil/news/story.asp?id=123164277.

Centers for Disease Control and Prevention. "Researchers Reconstruct 1918 Pandemic Influenza Virus; Effort Designed to Advance Preparedness," October 5, 2005. http://www.cdc.gov/media/pressrel/r051005.htm.

Commission to Assess United States National Security Space Management and Organization. *Report of the Commission to Assess United States National Security Space Management and Organization,* January 11, 2001. http://www.space.gov/docs/fullreport.pdf.

Commission to Assess the Threat to the United States from Electromagnetic Pulse (EMP) Attack. *Report of the Commission to Assess the Threat to the United States from Electromagnetic Pulse (EMP) Attack,* Volume 1: Executive Report, 2004. http://www.empcommission.org/docs/empc_exec_rpt.pdf.

Committee on Directed Energy Technology for Countering Indirect Weapons, National Research Council. *Review of Directed Energy Technology for Countering Rockets, Artillery, and Mortars (RAM): Abbreviated Version.* Washington, D.C.: The National Academies Press, 2008.

Committee on Implications of Emerging Micro- and Nanotechnologies, National Research Council. *Implications of Emerging Micro and Nanotechnology.* Washington, D.C.: The National Academies Press, 2002.

Committee on Military and Intelligence Methodology for Emergent Neurophysiological and Cognitive/Neural Research in the Next Two Decades, National Research Council. *Emerging Cognitive Neuroscience and Related Technologies.* Washington: The National Academy Press, 2008. http://www.nap.edu/catalog/12177.html.

Committee on Opportunities in Biotechnology for Future Army Applications, Board on Army Science and Technology, National Research Council. *Opportunities in Biotechnology for Future Army Applications.* Washington, D.C.: The National Academies Press, 2001.

Congressional Budget Office. *Energy Security in the United States,* May 9, 2012, http://www.cbo.gov/sites/default/files/cbofiles/attachments/05–09-Energy Security.pdf.

Congressional Budget Office. "The Veterans Health Administration's Treatment of PTSD and Traumatic Brain Injury among Recent Combat Veterans," February 9, 2012. http://www.cbo.gov/publication/42969.

Convention on Certain Conventional Weapons, 1980, http://treaties.un.org/Pages/ViewDetails.aspx?src=TREATY&mtdsg_no=XXVI-2&chapter=26&lang=en.

Convention on the Prohibition of the Use, Stockpiling, Production and Transfer of Anti-Personnel Mines and on their Destruction, September 18, 1997. http://www.icrc.org/ihl.nsf/FULL/580?OpenDocument.

Defense Advanced Research Product Agency. "A Huge Leap Forward in Robotics R&D: $2 Million Cash Prize Awarded to Stanford's 'Stanley' as Five Autonomous Ground Vehicles Complete DARPA Grand Challenge Course," October 9, 2005. http://archive.darpa.mil/grandchallenge05/gcorg/downloads/GC05%20Winner.pdf.

Defense Advanced Research Product Agency. "About." http://www.darpa.mil/About.aspx.

Defense Advanced Research Product Agency. "Adaptive Vehicle Make (AVM)." http://www.darpa.mil/AVM.aspx.

Defense Advanced Research Product Agency. "Blood Pharming." http://www.darpa.mil/Our_Work/DSO/Programs/Blood_Pharming.aspx.

Defense Advanced Research Product Agency. "Darpa Researchers Design Eye-Enhancing Virtual Reality Contact Lenses," January 31, 2012, http://www.darpa.mil/NewsEvents/Releases/2012/01/31.aspx.

Defense Advanced Research Product Agency. "Deep Learning." http://www.darpa.mil/Our_Work/I2O/Programs/Deep_Learning.aspx.

Defense Advanced Research Product Agency. "Education Dominance." http://www.darpa.mil/Our_Work/DSO/Programs/Education_Dominance.aspx.

Defense Advanced Research Product Agency. "Overview of the DARPA Augmented Cognition Technical Integration Experiment," May 8, 2008. http://www.dtic.mil/cgi-bin/GetTRDoc?AD=ADA475406.

Defense Advanced Research Product Agency. "Urban Challenge." http://archive.darpa.mil/grandchallenge/index.asp.

Defense Advanced Research Product Agency. "Z-Man." http://www.darpa.mil/Our_Work/DSO/Programs/Z_Man.aspx.

Defense Science Board. *Report of Defense Science Board Task Force on Directed Energy Weapons*, December 2007. http://www.dtic.mil/cgi-bin/GetTRDoc?AD=ADA476320.

Department of the Air Force. "Counter-Electronics High Power Microwave Advanced Missile Project (CHAMP) Joint Capability Technology Demonstration (JCTD), October 18, 2008. https://www.fbo.gov/index?s=opportunity&mode=form&id=e2daa9dccf59c9887810286dc9909d54&tab=core&_cview=1.

Department of Defense. "Arming Soldiers with Nutrition," *The Warrior*, January-February 2000. http://www.natick.army.mil/about/pao/pubs/warrior/00/janfeb/patch.htm.

Department of Defense. *Defense Advanced Research Projects Agency Justification Book Volume 1 Research, Development, Test & Evaluation, Defense-Wide Fiscal Year (FY) 2012 Budget Estimates.* www.darpa.mil/WorkArea/DownloadAsset.aspx?id=2400.

Department of Defense. *Defense Nanotechnology Research and Development Program: Report to Congress*, December 1, 2009. http://nano.gov/node/621.

Department of Defense. *Department of Defense Strategy for Operating in Cyberspace*, July 2011. http://www.defense.gov/news/d20110714cyber.pdf.

Department of Defense. "DoD News Briefing with Secretary Gates from the Pentagon," April 6, 2009. http://www.defense.gov/transcripts/transcript.aspx?transcriptid=4396.

Department of Defense. *Field Manual 90–44/6–22.5 Combat Stress.* https://rdl.train.army.mil/catalog/view/100.ATSC/AF18AF6D-7DFB-4CA4-BE78-8D0A5F EDF993–1274316004566/6–22.5/TOC.HTM.

Department of Defense. "Findings of the Nuclear Posture Review," January 9, 2002. http://www.defenselink.mil/news/Jan2002/g020109-D-6570C.html.

Department of Defense. *FY2009–2034 Unmanned Systems Integration Roadmap*, 2009. http://www.acq.osd.mil/psa/docs/UMSIntegratedRoadmap2009.pdf.

Department of Defense. *Military and Security Developments Involving the People's Republic of China 2010.* http://www.defense.gov/pubs/pdfs/2010_CMPR_Final.pdf.

Department of Defense. *Operational Rations of the Department of Defense.* http://nsrdec.natick.army.mil/media/print/OP_Rations.pdf.

Department of Defense. "Performance Optimizing Ration Components." http://nsrdec.natick.army.mil/media/fact/food/perc.pdf.

Department of Defense. "Probiotics, Prebiotics, and Keif." http://nsrdec.natick.army.mil/media/fact/food/Probiotics.pdf.

Department of Defense. *Report of Defense Science Board Task Force on High Energy Laser Weapon Systems Application*, June 2001. http://www.acq.osd.mil/dsb/reports/rephel.pdf.

Department of Defense Non-Lethal Weapons Program. "Active Denial System Fact Sheet." http://jnlwp.defense.gov/pressroom/adt.html.

Department of Energy, Advanced Research Projects Agency—Energy. "About." http://arpa-e.energy.gov/About/About.aspx.

Department of Energy Office of Science. "Artificial Retina Project." http://artificialretina.energy.gov/index.shtml.

Department of Energy. *Human Genome Project Information.* http://www.ornl.gov/sci/techresources/Human_Genome/home.shtml.

"Draft Treaty for the Prevention of Placement of Weapons in Outer Space," February 12, 2008. http://www.ln.mid.ru/brp_4.nsf/e78a48070f128a7b43256999005bcbb3/0d6e0c64d34f8cfac32573ee002d082a?OpenDocument.

European Science Foundation. "Big Molecules Join Together Will Lead to Better Drugs, Workshop Found," February 20, 2008. http://www.esf.org/media-centre/ext-single-news/article/new-understanding-of-how-big-molecules-join-together-will-lead-to-better-drugs-synthetic-organic-ma.html.

Federal Aviation Administration. *The Effects of Laser Illumination on Operational and Visual Performance of Pilots During Final Approach*, June 2004. http://www.faa.gov/library/reports/medical/oamtechreports/2000s/media/0409.pdf.

Federal Bureau of Investigation. "Amerithrax or Anthrax Investigation." http://www.fbi.gov/about-us/history/famous-cases/anthrax-amerithrax/amerithrax-investigation.

Garamone, Jim. American Forces Press Service. "CENTCOM Charts Operation Iraqi Freedom Progress," March 25, 2003. http://www.defenselink.mil/news/newsarticle.aspx?id=29230.

Garamone, Jim. "Patch May Deliver Nutrients to Future Warfighters," *American Forces Press Service*, February 28, 2000. http://www.defense.gov/news/newsarticle.aspx?id=44529.

General Accounting Office. *Assessments of Selected Major Weapon Programs*, March 2005, http://www.gao.gov/new.items/d05301.pdf.

Government Accountability Office. *DOD Efforts to Develop Laser Weapons for Theater Defense*, March 1999, http://www.gao.gov/products/NSIAD-99–50.

Jean, Grace, Office of Naval Research. "With a Bang, Navy Begins Tests on Electromagnetic Railgun Prototype Launcher," February 28, 2012. http://www.onr.navy.mil/Media-Center/Press-Releases/2012/Electromagnetic-Railgun-BAE-Prototype-Launcher.aspx.

Kan, Shirley. *China's Anti-Satellite Weapon Test, Congressional Research Service Report for Congress*, April 23, 2007, http://www.dtic.mil/cgi-bin/GetTRDoc?AD=AD A468025&Location=U2&doc=GetTRDoc.pdf.

Kaszav. Browner, January 8, 1998. http://archive.ca9.uscourts.gov/ca9/newopinions. nsf/04485f8dcbd4e1ea882569520074e698/54eb6df18826949b88256e5a 007188b2?OpenDocument.

Kenny, John M., et al. *A Narrative Summary and Independent Assessment of the Active Denial System, The Human Effects Advisory Panel*, February 2011. http://jnlwp. defense.gov/pdf/heap.pdf.

Lopez, C. Todd. "PEO Soldier Showcases Gear at Pentagon," *Army News Service*, June 16, 2009. http://www.army.mil/article/22755/peo-soldier-showcases-gear-at-pentagon/?ref=news-arnews-title1.

McCluney, Christen N. "'First-Strike Ration' Aims for Better Nutrition," *American Forces Press Service*, November 25, 2009. http://www.defense.gov/news/newsarticle.aspx?id=56854.

Missile Defense Agency. "Airborne Laser Test Bed Successful in Lethal Intercept Experiment," February 11, 2010. http://www.mda.mil/news/10news0002.html.

National Aeronautics and Space Administration. "Audacious & Outrageous: Space Elevators," September 7, 2000. http://science.nasa.gov/science-news/science-at-nasa/2000/ast07sep_1/.

National Aeronautics and Space Administration. "NASA Reaches Milestone in Space Launch Initiative Program; Also Announces No SLI Funding for X-33 or X-34," March 1, 2001. http://www.nasa.gov/home/hqnews/2001/01–031.txt.

National Aeronautics and Space Administration. "NASA—X-37 fact sheet (05/03)," June 2003. http://www.nasa.gov/centers/marshall/news/background/facts/x37facts2.html.

National Aeronautics and Space Administration. "NASA's X-43A Scramjet Breaks Speed Record," November 16, 2004. http://www.nasa.gov/missions/research/x43_schedule.html.

National Aeronautics and Space Administration. "STS-400: Ready and Waiting," May 5, 2009. http://www.nasa.gov/audience/foreducators/sts400-ready-and-waiting.html.

National Aeronautics and Space Administration. "X-15—Hypersonic Research at the Edge of Space," February 24, 2000. http://history.nasa.gov/x15/cover.html.

National Defense Authorization Fiscal Year 2001. http://www.dod.mil/dodgc/olc/docs/2001NDAA.pdf.

National Institute for Occupational Safety and Health, Centers for Disease Control and Prevention. *Approaches to Safe Nanotechnology,* 2009. http://www.cdc.gov/niosh/docs/2009–125/.

National Institute for Occupational Safety and Health, Centers for Disease Control and Prevention. "Nanotechnology." http://www.cdc.gov/niosh/topics/nanotech/.

National Institute of Standards and Technology. "CODATA Value: Speed of Light in Vacuum." http://physics.nist.gov/cgi-bin/cuu/Value?c.

National Institute of Standards and Technology. "Dietary Supplement Analysis Quality Assurance Program." http://www.nist.gov/mml/analytical/organic/dsqap.cfm.

National Museum of the US Air Force. "Fact Sheet: Boeing LGM-118A Peacekeeper." http://www.nationalmuseum.af.mil/factsheets/factsheet.asp?id=12225.

National Museum of the US Air Force. "Fact Sheet: Lockheed B-71 (SR-71)." http://www.nationalmuseum.af.mil/factsheets/factsheet.asp?id=2699.

National Nanotechnology Initiative. "NSET's Participating Federal Partners." http://nano.gov/partners.

The National Nanotechnology Initiative, Supplement to the President's FY 2012 Budget. http://www.nano.gov/sites/default/files/pub_resource/nni_2012_budget_supplement.pdf.

National Science and Technology Council Committee on Technology, Subcommittee on Nanoscale Science, Engineering, and Technology, National Nanotechnology Initiative. *Environmental, Health, and Safety Research Strategy,* October 2011. http://www.nano.gov/sites/default/files/pub_resource/nni_2011_ehs_research_strategy.pdf.

National Science Foundation. "Edible Vaccinations." http://www.nsf.gov/od/lpa/nsf50/nsfoutreach/htm/n50_z2/pages_z3/16_pg.htm.

National Space Policy of the United States of America, June 28, 2010. http://www.whitehouse.gov/sites/default/files/national_space_policy_6–28–10.pdf.

North Atlantic Treaty Organization. *179 STCME 05 E—The Security Implications of Nanotechnology.* http://www.nato-pa.int/default.asp?SHORTCUT=677.

Presidential Determination on Classified Information Concerning the Air Force's Operating Location Near Groom Lake, Nevada, No. 95–45 of September 29, 1995. http://frwebgate.access.gpo.gov/cgi-bin/getdoc.cgi?dbname=1995_register&docid=fr10oc95–127.pdf.

Report to the President and Congress on the Third Assessment of the National Nanotechnology Initiative, March 12, 2010. http://www.whitehouse.gov/sites/default/files/microsites/ostp/pcast-nano-report.pdf.

Richardson, B. S., editor. *Phase I Report: DARPA EXOSKELETON PROGRAM.* Oak Ridge, TN: Oak Ridge National Laboratory, 2004. http://www.ornl.gov/~webworks/cppr/y2004/rpt/118274.pdf.

Rupp, Sheila. "Operationally Responsive Space," *Air Force Print News*, May 22, 2007, http://www.kirtland.af.mil/news/story.asp?id=123054292.

Russian Federation. "On Presentation of the World Citizens Award to Stanislav Petrov," January 19, 2006. http://www.un.int/russia/other/060119eprel.pdf.

Sulsky, Sandra I., John D. Grabenstein, and Rachel Gross Delbos. "Disability among U.S. Army Personnel Vaccinated Against Anthrax," *Journal of Occupational and Environmental Medicine*, Volume 46, Number 10, October 2004. http://www.anthrax.mil/documents/library/Anthrax2004.pdf.

United Kingdom, Department for Business Innovation & Skills. *A.I. Law: Ethical and Legal Dimensions of Artificial Intelligence*, October 5, 2011. http://www.sigmascan.org/Live/Issue/ViewIssue/485/1/a-i-law-ethical-and-legal-dimensions-of-artificial-intelligence/.

United States. "Iris Recognition," August 2007. http://www.biometrics.gov/Documents/irisrec.pdf.

United States. *The National Strategy to Secure Cyberspace*. February 2003.

United States Air Force. "Factsheet: JOINT DIRECT ATTACK MUNITION GBU-31/32/38," November 2007. http://www.af.mil/factsheets/factsheet.asp?id=108.

United States Air Force. "Factsheets: Advanced Extremely High Frequency (AEHF) System." http://www.losangeles.af.mil/library/factsheets/factsheet.asp?id=5319.

United States Air Force. "Factsheets: B-52 STRATOFORTRESS." http://www.af.mil/information/factsheets/factsheet.asp?id=83.

United States Air Force. "Factsheets: X-37B Orbital Test Vehicle," March 3, 2011. http://www.af.mil/information/factsheets/factsheet.asp?fsID=16639.

United States Air Force. "Program Elements FY2009—Counterspace Systems," February 2008. http://www.js.pentagon.mil/descriptivesum/Y2009/AirForce/0604421F.pdf.

United States Air Force. *Report on Technology Horizons: A Vision for Air Force Science & Technology During 2010–2030*, Volume 1, May 15, 2010. http://www.af.mil/shared/media/document/AFD-100727–053.pdf.

United States Air Force, 30th Space Wing Public Affairs. "X-37B Orbital Test Vehicle lands at Vandenberg," June 16, 2012. http://www.vandenberg.af.mil/news/story.asp?id=123306242.

United States Army. "Connecting Soldiers to Digital Applications," *Stand-To!*, July 15, 2010. http://www.army.mil/standto/archive/2010/07/15/.

United States Internet Crime Complaint Center. *2010 Internet Crime Report*. http://www.ic3.gov/media/annualreports.aspx.

United States Navy. "First Navy Trainer Completes Biofuel Flight at Patuxent River," August 25, 2011. http://www.navy.mil/search/display.asp?story_id=62384.

United States Navy. *Naval S&T Strategic Plan*. http://www.onr.navy.mil/en/About-ONR/~/media/Files/About%20ONR/Naval-Strategic-Plan.ashx.

United States Navy. "Office of Naval Research Achieves Milestone With Free Electron Laser Program," January 19, 2011. http://www.onr.navy.mil/Media-Center/Press-Releases/2011/Free-Electron-Laser-Milestone.aspx.

United States Navy, Office of Naval Research. "Electromagnetic Railgun." http://www.onr.navy.mil/en/Science-Technology/Departments/Code-35/All-Programs/air-warfare-352/Electromagnetic-Railgun.aspx.

Young, Jim. Air Force Flight Test Center History Office. "Milestones in Aerospace History at Edwards AFB," June 2011. http://www.af.mil/shared/media/document/AFD-080123–063.pdf.

Secondary

Ackerman, Spencer. "Navy's Rail Gun Blasts through Budget Restrictions," *Wired*, February 10, 2012. http://www.wired.com/dangerroom/2012/02/rail-gun/.

Adee, Sally. "Winner: The Revolution Will Be Prosthetized," *IEEE Spectrum*, January 2009. http://spectrum.ieee.org/robotics/medical-robots/winner-the-revolution-will-be-prosthetized.

Amnesty International. *'Less Than Lethal'? The Use of Stun Weapons in US Law Enforcement*, 2008. http://www.amnesty.org/en/library/info/AMR51/010/2008/en.

Arms Control Association. "U.S. Test-Fires 'MIRACL' at Satellite Reigniting ASAT Weapons Debate," *Arms Control Today*, October 1997. http://www.armscontrol.org/act/1997_10/miracloct.

Arthur, W. Brian. *The Nature of Technology*. New York: Free Press, 2009.

Associated Press. "Army Bans Use of Privately Bought Armor," March 30, 2006. http://www.usatoday.com/news/washington/2006–03–30-bodyarmor_x.htm.

Astronautix. "Gnom." http://astronautix.com/lvs/gnom.htm.

Axe, David. "Combat Exoskeleton Marches toward Afghanistan Deployment," *Wired*, May 23, 2012. http://www.wired.com/dangerroom/2012/05/combatexoskeleton-afghanistan/#more-81313.

Axe, David. "New 'ICBMs vs. Terrorists' Plan: Now 50% Less Crazy!" *Wired*, March 7, 2011. http://www.wired.com/dangerroom/2011/03/new-icbms-vs-terrorists-plan-now-50-less-crazy/.

Axe, David. "Semper Fly: Marines in Space." *Popular Science*, December 2006. http://www.popsci.com/military-aviation-space/article/2006–12/semper-fly-marines-space.

Beason, Doug. *The E-Bomb*. Cambridge, Massachusetts: Da Capo Press, 2005.

Berger, Brian and Amy Klamper. "NASA Propulsion Plans Resonate with Some in Rocket Industry," *Space News*, February 26, 2010. http://www.spacenews.com/launch/100226-nasa-propulsion-plans-resonate-rocket-industry.html.

Berkeley Robotics & Human Engineering Laboratory. "BLEEX." http://bleex.me.berkeley.edu/research/exoskeleton/bleex/.

Berkowitz, Bruce. *The New Face of War: How War Will be Fought in the 21st Century*. New York: The Free Press, 2003.

Bernstein, Joseph A. "Seeing Crime Before it Happens," *Discovery*, January 23, 2012. http://discovermagazine.com/2011/dec/02-big-idea-seeing-crime-before-it-happens.

Blackmore, Tim. *War X: Human Extensions in Battlespace*. Toronto: University of Toronto Press, 2005.

Bloomberg Businessweek. "The Secret to Google's Success," March 6, 2006. http://www.businessweek.com/magazine/content/06_10/b3974071.htm.

Boeing. "Boeing Advanced Tactical Laser Defeats Ground Target in Flight Test," September 1, 2009. http://boeing.mediaroom.com/index.php?s=43&item=817.

Boeing. "Boeing Awarded Contract to Develop Counter-Electronics HPM Aerial Demonstrator," May 15, 2009. http://boeing.mediaroom.com/index.php?s=43&item=656.

Booth, Max. *War Made New: Weapons, Warriors, and the Making of the Modern World*. New York: Gotham, 2007.

Bowman, Tom. "China Protests after U.S. Shoots Down Satellite," *National Public Radio*, February 21, 2008. http://www.npr.org/templates/story/story.php?storyId=19246330.

Boyle, Rebecca. "How Modified Worms and Goats Can Mass-Produce Nature's Toughest Fiber," *Popular Science*, October 6, 2010. http://www.popsci.com/science/article/2010–10/fabrics-spider-silk-get-closer-reality.

Branaby, Frank and Marlies ter Borg, editors. *Emerging Technologies and Military Doctrine: A Political Assessment*. New York: St. Martin's Press, 1986.

British Broadcasting Corporation. "Bin Laden Compound Location Suggested by 2008 Study," May 3, 2011. http://www.bbc.co.uk/news/world-13275104.

British Broadcasting Corporation. "Calls to Ban 'Anti-Teen' Device," February 12, 2008. http://news.bbc.co.uk/2/hi/uk_news/7240180.stm.

British Broadcasting Corporation. "Measles Outbreak Fears Spread," January 4, 2002. http://news.bbc.co.uk/2/hi/health/1742177.stm.

British Broadcasting Corporation. "Stuxnet Worm Hits Iran Nuclear Plant Staff Computers," September 26, 2010. http://www.bbc.co.uk/news/world-middle-east-11414483.

British Broadcasting Corporation. "Tanks Test Infrared Invisibility Cloak," September 5, 2011. http://www.bbc.com/news/technology-14788009.

Broad, William J. "U.S. and Israel Shelved Laser as a Defense." *New York Times*, July 30, 2006, http://www.nytimes.com/2006/07/30/world/middleeast/30laser.html?_r=1.

Brooks, Rodney. *Flesh and Machines: How Robots Will Change Us*. New York: Pantheon Books, 2002.

Bullis, Kevin. "Climbing Walls with Carbon Nanotubes," *Technology Review*, June 25, 2007. http://www.technologyreview.com/computing/18966/?mod=related.

Cameron, Nigel M. de S. and M. Ellen Mitchell, editors. *Nanoscale: Issues and Perspectives for the Nano Century*. New Jersey: John Wiley & Sons, Inc., 2007.

Canadian Broadcasting Corporation. "Go-Pills, Bombs & Friendly Fire," November 17, 2004. http://www.cbc.ca/news/background/friendlyfire/gopills.html.

CBS Broadcasting Inc. "Obama on Bin Laden: The Full "60 Minutes" Interview," May 8, 2011. http://www.cbsnews.com/8301–504803_162–20060530–10391709.html.

Clark, Stephen. "Air Force Spaceplane Is an Odd Bird with a Twisted Past," April 2, 2010. http://spaceflightnow.com/atlas/av012/100402x37update/.

Clarke, Arthur C. *2010: Odyssey Two*. New York: Del Rey, 1982.

Coker, Christopher. *Ethics and War in the 21st Century*. New York: Routledge, 2008.

Collins, John M. *Military Space Forces*. Washington, DC: Pergamon-Braseesey's International Defense Publishers, 1989.

Cox, Matthew. "SF Units to Get Latest Land Warrior kit," *Army Times*, November 9, 2009. http://www.armytimes.com/news/2009/11/army_landwarrior_110909w/.

Czysz, Paul and Claudio Bruno. *Future Spacecraft Propulsion Systems: Enabling Technologies for Space Exploration*. Berlin: Springer, 2006.

Da Costa, Gilbert. "Setback for Nigeria's Polio Fighters," *Time*, October 25, 2007. http://www.time.com/time/health/article/0,8599,1675423,00.html.

Deakin Research. "Medical "Smart Bomb" Possibly the Key to Eradicating Cancer Cells," July 2010. http://www.gsdm.com.au/newsletters/deakin/july10/7.html.

Delude, Cathryn M. Massachusetts Institute of Technology. "Computer Model Mimics Neural Processes in Object Recognition," February 23, 2007. http://web.mit.edu/press/2007/surveillance.html.

Dillow, Clay. "DARPA Invests in Megapixel Augmented-Reality Contact Lenses," *Popular Science*, February 1, 2012. http://www.popsci.com/technology/article/2012–02/video-nano-enhanced-contact-lens-makes-augmented-reality-more-realistic.

Dillow, Clay. "IBM's Digital Billboard Displays Individualized Ads by Reading the RFID Data in Your Wallet," *Popular Science*, August 2, 2010. http://www.popsci.com/technology/article/2010–08/ibms-new-digital-billboard-tailors-individual-ads-rfid-data-your-credit-card.

Drexler, K. Eric. *Unbounding the Future: The Nanotechnology Revolution*. Foresight Institute, 1983. http://www.foresight.org/UTF/download/unbound.pdf.

Dust Networks. "Company Background." http://www.dustnetworks.com/about/company_overview.

Enserink, Martin. "Grudgingly, Virologists Agree to Redact Details in Sensitive Flu Papers," *Science*, 20 December 2011. http://news.sciencemag.org/scienceinsider/2011/12/grudgingly-virologists-agree-to.html.

Eshel, David. "Technology Shortens the Kill Chain in Urban Combat," *Aviation Week—Defense Technology International*, March 28, 2008. http://www.aviationweek.com/aw/generic/story_generic.jsp?channel=dti&id=news/DTIKILL.

xml&headline=Technology%20Shortens%20the%20Kill%20Chain%20 in%20Urban%20Combat.

Eurofighter Consortium. "Direct Voice Input." http://www.eurofighter.com/media/ press-office/facts-sheet-mediakit/direct-voice-input.html.

Evans, Jon. "Manufacturing the Carbon Nanotube Market," *Chemistry World*, November 2007, Vol. 4, No. 11. http://www.rsc.org/chemistryworld/Issues/ 2007/November/ManufacturingCarbonNanotubeMarket.asp.

Evans, Nicholas D. *Military Gadgets: How Advanced Technology Is Transforming Today's Battlefield*. New York: Financial Times-Prentice Hall, 2004.

Faber, Scott. "Printing in 3-D," *Discovery*, September 1994. http://discovermagazine. com/1994/sep/printingin3d426/?searchterm=3d.

Farwell, Lawrence A. "Brain Fingerprinting: A Comprehensive Tutorial Review of Detection of Concealed Information with Event-Related Brain Potentials," *Cognitive Neurodynamics*, Volume 6, Number 2 (2012), 115–54. http://www. springerlink.com/content/7710950336312146/.

Fédération Aéronautique Internationale. "Aeromodelling and Spacemodelling Records." http://www.fai.org/ciam-records.

Fédération Aéronautique Internationale. "Powered Aeroplane World Records." http://www.fai.org/record-powered-aeroplanes.

Federation of American Scientists. "Airborne Laser." http://www.fas.org/spp/starwars/ program/abl.htm.

Ferster, Warren and Colin Clark. "NRO Confirms Chinese Laser Test Illuminated U.S. Spacecraft," *Space News*, October 3, 2006. http://www.spacenews.com/ archive/archive06/chinalaser_1002.html.

Filton. "The printed world," *The Economist*, February 10, 2011. http://www.economist. com/node/18114221?story_id=18114221.

Fisher, Franklin. "U.S. Says Apache Copters Were Targeted by Laser Weapons Near Korean DMZ," *Stars and Stripes*, May 14, 2003. http://www.stripes.com/ news/u-s-says-apache-copters-were-targeted-by-laser-weapons-near-korean- dmz-1.9753.

Fontaine, Scott. "X-37B Test Mission Called Big Accomplishment," *Defense News*, December 6, 2010. http://defensenews.com/story.php?i=5176376&c= AME&s=TOP.

Foresight Institute. "About Nanotechnology." http://www.foresight.org/nano/index. html.

Friedman, George. *The Next 100 Years: A Forecast for the 21st Century*. New York: Doubleday, 2009.

Friedman, George and Meredith Friedman. *The Future of War: Power, Technology and American World Dominance in the Twenty-first Century*. New York: St. Martin's Griffin, 1998.

Friedman, Norman. *Unmanned Combat Air Systems: A New Kind of Carrier Aviation.* Annapolis: Naval Institute Press, 2010.

Friedman, Thomas L. *The World is Flat: A Brief history of the Twenty-First Century.* New York: Farrar, Straus and Giroux, 2005.

Fulghum, David. "For Now JSF Will Not Embrace Electronic Attack," *Aviation Week,* January 23, 2012. http://www.aviationweek.com/Article.aspx?id=/article-xml/ AW_01_23_2012_p24–415796.xml.

General Dynamics. "Autonomous Robotics Programs Vetronics Technology Integration (VTI)." http://www.gdrs.com/robotics/programs/program.asp? UniqueID=12.

General Electric. "The Story Behind the Real 'Iron Man' Suit," November 23, 2010. http://www.gereports.com/the-story-behind-the-real-iron-man-suit/.

Gilbert, Gary, et al. "USAMRMC TATRC Combat Casualty Care and Combat Service Support Robotics Research & Technology Programs." http://www.tatrc. org/ports/robotics/docs/USAMRMC_CCC_CSS_Robotics.pdf.

Gillespie, Thomas W. and John A. Agnew. "Finding Osama Bin Laden: An Application of Biogeographic Theories and Satellite Imagery," *MIT International Review,* February 17, 2009. http://web.mit.edu/mitir/2009/online/finding-bin-laden.pdf.

Global Security. "Advanced Extremely High Frequency (AEHF)." http://www.glo balsecurity.org/space/systems/aehf.htm.

Global Security. "Dense Inert Metal Explosive (DIME)." http://www.globalsecurity. org/military/systems/munitions/dime.htm.

Global Security. "DF-21/CSS-5." http://www.globalsecurity.org/wmd/world/china/ df-21.htm.

Global Security. "Mobile Tactical High Energy Laser." http://www.globalsecurity.org/ space/systems/mthel.htm.

Goldberg, A., et al. "Dual-Band Imaging of Military Targets Using a QWIP Focal Plane Array." http://www.dtic.mil/cgi-bin/GetTRDoc?AD=ADA392953.

Goldstein, Seth Copen, Jason D. Campbell, and Todd C. Mowry. "Programmable Matter," *IEEE Computer* 38(6): 99–101, June 2005. http://www.cs.cmu. edu/~claytronics/papers/goldstein-computer05.pdf.

Golovanov, Sergey. "TDL4 Starts Using 0-Day Vulnerability!" *Securelist,* http://www. securelist.com/en/blog/337/TDL4_Starts_Using_0_Day_Vulnerability.

Goodman, D.R. "Nephrotoxicity. Toxic effects in the kidneys," in *Industrial Toxicology Safety and Health Applications in the Workplace,* ed. Williams, P.L. and Burson, J.L. (New York, NY: Van Nostrand Reinhold Company, 1985), 106–22. Quoted in CDC http://www.atsdr.cdc.gov/ToxProfiles/tp150-c3.pdf.

Gorman, Siobhan, Yochi J. Dreazen, and August Cole. "Insurgents Hack U.S. Drones," *Wall Street Journal,* December 17, 2009. http://online.wsj.com/article/ SB126102247889095011.html.

Graham-Rowe, Duncan. "'Gadget Printer' Promises Industrial Revolution," *New Scientist*, January 8, 2003. http://www.newscientist.com/article/dn3238.

Greenemeier, Larry. "National Robotics Week to Highlight the Past, Present and Future of Robot Research," *Scientific American*, February 9, 2010. http://www.scientificamerican.com/blog/post.cfm?id=national-robotics-week-to-highlight-2010–02–09.

Greenemeier, Larry. "Seeking Address: Why Cyber Attacks Are So Difficult to Trace Back to Hackers," *Scientific American*, June 11, 2011. http://www.scientificamerican.com/article.cfm?id=tracking-cyber-hackers&WT.mc_id=SA_Twitter_sciam.

Grossman, Dave. "The Myth of our Returning Veterans and Violent Crime," *Inside Homeland Security*, Spring 2011. http://killology.com/Myth%20of%20returning%20vets%20&%20violence.pdf.

Gruerer, Wolfang. "OCZ Launches Brain-Computer-Interface," *Tom's Hardware*, March 3, 2008. http://www.tomshardware.com/news/ocz-launches-brain-computer-interface,4897.html.

Guizzo, Erico and Harry Goldstein. "The Rise of the Body Bots," *IEEE Spectrum*, October 2005. http://spectrum.ieee.org/biomedical/bionics/the-rise-of-the-body-bots.

Guynup, Sharon. "Seeds of a New Medicine," *Genome News Network*, July 28, 2000. http://www.genomenewsnetwork.org/articles/07_00/vaccines_trees.shtml.

Hambling, David. "Cloak of Light Makes Drone Invisible," *Wired*, May 9, 2008. http://www.wired.com/dangerroom/2008/05/invisible-drone.

Hambling, David. "Microwave Weapon Will Rain Pain from the Sky," *New Scientist*, July 23, 2009. http://www.newscientist.com/article/mg20327185.600-microwave-weapon-will-rain-pain-from-the-sky.html.

Hambling, David. "US Boasts of Laser Weapon's 'Plausible Deniability'," *New Scientist*, August 12, 2008. http://www.newscientist.com/article/dn14520-us-boasts-of-laser-weapons-plausible-deniability.html.

Hambling, David. *Weapons Grade: How Modern Warfare Gave Birth to Our High-Tech World*. New York: Carroll &Graff Publishers, 2005.

Haney, Eric L. and Brian M. Thompsen, editors. *Beyond Shock and Awe: Warfare in the 21st Century*. New York: Berkley Caliber, 2006.

Harel, Amos and Avi Issacharoff. "Top Official: Israel Gave No Guarantees in Exchange for Gaza Truce," *Haaretz*, March 14, 2012, http://www.haaretz.com/print-edition/news/top-official-israel-gave-no-guarantees-in-exchange-for-gaza-truce-1.418328.

Harley, Naomi, et al. *A Review of the Scientific Literature As It Pertains to Gulf War Illnesses Volume 7: Depleted Uranium*. Santa Monica, CA: RAND Corporation, 1999. http://www.rand.org/pubs/monograph_reports/MR1018z7.html.

Hendrickx, Bart and Bert Vis. *Energiya-Buran: The Soviet Space Shuttle*. Chichester, UK: Springer, 2007.

Hentoff, Nat. CATO Institute. "Few Batting Eyes at Obama's Deadly Drone Policy," *Cato.org*, July 29, 2009. http://www.cato.org/pub_display.php?pub_id=12012.

Herman, Arthur. *To Rule the Waves*. New York: Harper Collins, 2004.

Hobbs, David. *An Illustrated Guide to Space Warfare*. New York: Prentice Hall, 1986.

Hoffman, David. "I Had A Funny Feeling in My Gut," *Washington Post*, February 10, 1999. http://www.washingtonpost.com/wp-srv/inatl/longterm/coldwar/shatter021099b.htm.

Holden, Michael. "Make Cupcakes, Not Bombs," *Reuters*, June 3, 2011. http://uk.reuters.com/article/2011/06/03/uk-britain-mi6-hackers-idUKLNE7520322011 0603.

Holder, William G. and William D. Siuru, Jr. "Some Thoughts on Reusable Launch Vehicles," *Air University Review*, November-December 1970. http://www.airpower.au.af.mil/airchronicles/aureview/1970/nov-dec/holder.html

Howard, Sean. "Nanotechnology and Mass Destruction: The Need for an Inner Space Treaty," *Disarmament Diplomacy*, No. 65, July–August 2002. http://www.acronym.org.uk/dd/dd65/65op1.htm.

Hsiao, Russell. "Aims and Motives of China's Recent Missile Defense Test," *China Brief*, Volume 10, Issue 2, January 21, 2010, http://www.jamestown.org/single/?no_cache=1&tx_ttnews[tt_news]=35943.

Hudson, Alex. "Is Graphene a Miracle Material?" *BBC*, May 21, 2011. http://news.bbc.co.uk/2/hi/programmes/click_online/9491789.stm.

IBM. "IBM Spelled with 35 Xenon Atoms," September 28, 2009. http://www-03.ibm.com/press/us/en/photo/28500.wss.

IEEE Global History Network. "Theodore H. Maiman." http://www.ieeeghn.org/wiki/index.php/Theodore_H._Maiman.

"IGEM/Learn About," September 26, 2011. http://igem.org/wiki/index.php?title=IGEM/Learn_About&oldid=4750.

Imai, Masaki, et al. "Experimental Adaptation of an Influenza H5 HA Confers Respiratory Droplet Transmission to a Reassortant H5 HA/H1N1 Virus in Ferrets," *Nature*, May 22, 2012. http://www.nature.com/nature/journal/vaop/ncurrent/full/nature10831.html.

Intel. "Moore's Law and Intel Innovation." http://www.intel.com/about/companyinfo/museum/exhibits/moore.htm.

International Risk Governance Council. *Nanotechnology Risk Governance*, 2007. http://www.irgc.org/IMG/pdf/PB_nanoFINAL2_2_.pdf.

Jane's Information Group. "Laser Weapons (China), Defensive Weapons." http://www.janes.com/articles/Janes-Strategic-Weapon-Systems/Laser-weapons-China.html.

Jones, Richard A. L. *Soft Machines: Nanotechnology and Life*. New York: Oxford University Press, 2004.

Joy, Bill. "Why the Future Doesn't Need Us," *Wired*, Issue 8.4, April 2000. http://www.wired.com/wired/archive/8.04/joy.html.

Kaku, Michio. *The Physics of the Future*. New York: Double Day, 2011.

Kanellos, Michael. "Moore's Law to Roll on for Another Decade," *CNET*, February 10, 2003. http://news.cnet.com/2100–1001–984051.html.

Keim, Brandon. "Uncle Sam Wants Your Brain," *Wired*, August 13, 2008. http://www.wired.com/wiredscience/2008/08/uncle-sam-wants/.

Kenyon, Henry S. "Programmable Matter Research Solidifies," *Signal*, June 2009. http://www.afcea.org/signal/articles/templates/Signal_Article_Template.asp?articleid=1964.

Knight, Helen. "E. coli Engineered to Make Convenient 'Drop-In' Biofuel," *New Scientist*, July 29, 2010. http://www.newscientist.com/article/dn19238-e-coli-engineered-to-make-convenient-dropin-biofuel.html.

Koblentz, Gregory D. *Living Weapons*. Ithaca: Cornell University Press, 2009.

Koerner, Brendan I. "What Is Smart Dust, Anyway?" *Wired*, June 2003, http://www.wired.com/wired/archive/11.06/start.html?pg=10.

Koplow, David A. *Death by Moderation: The U.S. Military's Quest for Useable Weapons*. New York: Cambridge University Press, 2010.

Kopp, Carlo. "The Electromagnetic Bomb—A Weapon of Electrical Mass Destruction,"*Air & Space Power Journal*, 1996. http://www.airpower.maxwell.af.mil/airchronicles/cc/apjemp.html.

Kurzweil, Ray. "How My Predictions Are Faring," October 2010, http://www.kurzweilai.net/predictions/download.php.

Kurzweil, Ray. *The Age of Spiritual Machines*. New York: Viking, 1999.

La Franchi, Peter. "Iranian-Made Ababil-T Hezbollah UAV Shot Down by Israeli Fighter in Lebanon Crisis," *Flight International*, August 15, 2006. http://www.flightglobal.com/articles/2006/08/15/208400/iranian-made-ababil-t-hezbollah-uav-shot-down-by-israeli-fighter-in-lebanon.html.

La Franchi, Peter. "Near Misses between UAVs and Airliners Prompt NATO Low-Level Rules Review," *Flight International*, March 14, 2006. http://www.flightglobal.com/articles/2006/03/14/205379/animation-near-misses-between-uavs-and-airliners-prompt-nato-low-level-rules.html.

Lacey, Marc. "Look at the Place! Sudan Says, 'Say Sorry', but U.S. Won't," *New York Times*, October 20, 2005. http://www.nytimes.com/2005/10/20/international/africa/20khartoum.html?_r=1.

Langston, Lee S. "Crown Jewels: These Crystals Are the Gems of Turbine Efficiency," *Mechanical Engineering Magazine*, February 2006. http://memagazine.asme.org/Articles/2006/February/Crown_Jewels.cfm.

Leiss, William. "Policing Science: Genetics, Nanotechnology, Robotics," *Technikfolgenabschätzung—Theorie und Praxis*, December 2004, Vol. 13, No. 3. http://www.itas.fzk.de/tatup/043/inhalt.htm.

Lele, Ajey. *Strategic Technologies for the Military: Breaking New Frontiers*. New Delhi: SAGE Publications India Pvt. Ltd., 2009.

Levinson, Charles. "Israeli Robots Remake Battlefield," *Wall Street Journal*, January 13, 2010. http://online.wsj.com/article/SB126325146524725387.html#MARK.

Lieber, Keir A. *War and the Engineers*. Ithaca: Cornell University Press, 2005.

Lockheed Martin. "Advanced Extremely High Frequency (AEHF)." http://www.lockheedmartin.com/us/products/advanced-extremely-high-frequency—aehf-.html.

Lupton, David E. *On Space Warfare*. Maxwell Air Force Base, AL: Air University Press, 1998. http://www.airpower.maxwell.af.mil/airchronicles/apj/apj98/win98/deblois.html.

Lusaka, Olga Manda. "Controversy Rages over 'GM' Food Aid," *Africa Renewal*, Vol. 16 #4, February 2003. http://www.un.org/en/africarenewal/vol16no4/164food2.htm.

Madrigal, Alexis. "Scientists Build World's Smallest Transistor, Gordon Moore Sighs with Relief," *Wired*, April 17, 2008. http://www.wired.com/wiredscience/2008/04/scientists-buil/.

Makar, Jon, Jim Margeson, and Jeanne Luh. *Carbon Nanotube/Cement Composites—Early Results and Potential Applications*. http://www.nrc-cnrc.gc.ca/obj/irc/doc/pubs/nrcc47643/nrcc47643.pdf.

Markman, Art. "Are ADHD Drugs Smart Pills?" *Psychology Today*, September 6, 2011. http://www.psychologytoday.com/blog/ulterior-motives/201109/are-adhd-drugs-smart-pills.

Markoff, John. "Taking Spying to Higher Level, Agencies Look for More Ways to Mine Data," *New York Times*, February 25, 2006. http://www.nytimes.com/2006/02/25/technology/25data.html?ref=johnmarkoff&pagewanted=print.

Marks, Paul. "3D Printing: The World's First Printed Plane," *New Scientist*, August 1, 2011. http://www.newscientist.com/article/dn20737-3d-printing-the-worlds-first-printed-plane.html?full=true.

Martel, William, editor. *Technological Arsenal*. Washington: Smithsonian, 2001.

Massachusetts Institute of Technology Institute for Soldier Nanotechnologies. "Project 2.3.2: Non-Invasive Delivery and Sensing." http://web.mit.edu/isn/research/sra02/project02_03_02.html.

Massie, Robert K. *Dreadnought: Britain, Germany, and the Coming of the Great War*. New York: Ballantine Books, 1991.

Mazzetti, Mark, Helene Cooper, and Peter Baker. "Behind the Hunt for Bin Laden," *New York Times*, May 2, 2011. http://www.nytimes.com/2011/05/03/world/asia/03intel.html?_r=1.

McDonald, Avril. "Depleted Uranium Weapons: The Next Target for Disarmament?" *Disarmament Forum*, Vol. three, 2008. http://www.unidir.org/pdf/articles/pdf-art2757.pdf.

McIntosh, Daniel. "The Transhuman Security Dilemma," *Journal of Evolution & Technology*, Vol. 21, Issue 2, December 2010. http://jetpress.org/v21/mcintosh.htm.

McKenna, Phil. "A New Route to Cellulosic Biofuels," *Technology Review*, November 20, 2009. http://www.technologyreview.com/energy/23989/.

McLeary, Paul, Sharon Weinberger, and Angus Batey. "Drone Impact on Pace of War Draws Scrutiny," *Aviation Week*, July 8, 2011. http://www.aviationweek.com/aw/generic/story_generic.jsp?channel=dti&id=news/dti/2011/07/01/DT_07_01_2011_p40–337605.xml&headline=Drone%20Impact%20On%20Pace%20Of%20War%20Draws%20Scrutiny.

McLellan, Tom M., et al.. "The Impact of Caffeine on Cognitive and Physical Performance and Marksmanship during Sustained Operations," *Canadian Military Journal*, Winter 2003–2004. http://www.journal.forces.gc.ca/vo4/no4/military-meds-eng.asp.

McLoughlin, Michael P. "DARPA Revolutionizing Prosthetics 2009," January 2009. http://www.dtic.mil/cgi-bin/GetTRDoc?Location=U2&doc=GetTRDoc.pdf&AD=ADA519193.

Milian, Mark. "Facebook Lets Users Opt Out of Facial Recognition," *CNN*, June 9, 2011. http://www.cnn.com/2011/TECH/social.media/06/07/facebook.facial.recognition/index.html?iref=allsearch.

Milian, Mark. "Google Taking More Cautious Stance on Privacy," *CNN*, June 1, 2011. http://www.cnn.com/2011/TECH/web/05/31/google.schmidt/index.html.

Miller, Greg. "Bin Laden Files Show Al-Qaeda under Pressure," *Washington Post*, July 1, 2011. http://www.washingtonpost.com/national/national-security/bin-laden-document-trove-reveals-strain-on-al-qaeda/2011/07/01/AGDN4luH_print.html.

Mims, Christopher. "Microwave-Powered Rocket Ascends without Fuel," *Technology Review*, September 7, 2010. http://www.technologyreview.com/blog/mimssbits/25701/.

Mooney, Chris. "Why Does the Vaccine/Autism Controversy Live On?" *Discover*, May 6, 2009. http://discovermagazine.com/2009/jun/06-why-does-vaccine-autism-controversy-live-on/article_view?b_start:int=3&-C=.

Mowthorpe, Matthew. *The Militarization and Weaponization of Space*. Toronto, ON: Lexington Books, 2004.

Mueller, Karl P. "Totem and Taboo: Depolarizing the Space Weaponization Debate." Paper based on presentation given to Weaponization of Space Project of the Eliot School of International Affairs Space Policy Institute and Security Policy Studies Program, George Washington University, December 3, 2001. http://www.gwu.edu/~spi/spaceforum/TotemandTabooGWUpaperRevised%5B1%5D.pdf.

Mulhal, Douglas. *Our Molecular Future—How Nanotechnology, Robotics, Genetics, and Artificial Intelligence Will Transform Our World.* Amherst: Prometheus Books, 2002.

Mullard, Asher. "Mediator Scandal Rocks French Medical Community," *The Lancet,* Vol. 377, March 12, 2011. http://download.thelancet.com/pdfs/journals/lancet/PIIS0140673611603346.pdf.

Myrabo, Leik N. and Donald G. Messittf. "Ground and Flight Tests of a Laser Propelled Vehicle," 1997. http://pdf.aiaa.org/downloads/1998/1998_1001.pdf?CFID=1326408&CFTOKEN=88988801&.

Neufeld, Michael J. *Von Braun: Dreamer of Space, Engineer of War.* New York: Vintage Books, 2008.

The Nobel Foundation. "Charles H. Townes—Biography." http://www.nobelprize.org/nobel_prizes/physics/laureates/1964/townes-bio.html.

Northrop Grumman. "Northrop Grumman Chosen to Increase Efficiency for Next-Generation Military Laser Technology," September 28, 2010. http://www.irconnect.com/noc/press/pages/news_releases.html?d=202483.

Northrop Grumman. "Northrop Grumman Laser 'Firsts.'" http://www.as.northropgrumman.com/by_capability/directedenergy/laserfirsts/index.html

Oberg, James. "Astronaut." World Book Online Reference Center. Chicago, IL: World Book, Inc., 2005. http://www.worldbookonline.com/wb/Article?id=ar034800.

O'Mathúna, Dónal P. *Nanoethics: Big Ethical Issues with Small Technology.* United Kingdom: Continuum, 2009.

Osborn, Kris. "FCS Is Dead; Programs Live On," *Defense News,* May 18 2009. http://www.defensenews.com/story.php?i=4094484.

Palfrey, John. "Middle East Conflict and an Internet Tipping Point," *Technology Review,* February 25, 2011. http://www.technologyreview.com/web/32437/?mod=chthumb.

Paschotta, Rüdiger. "Beam Quality," *Encyclopedia for Photonics and Laser Technology.* http://www.rp-photonics.com/beam_quality.html.

Penenberg, Adam L. "The Surveillance Society," *Wired,* Issue 9.12, December 2001. http://www.wired.com/wired/archive/9.12/surveillance.html.

Pennisi, Elizabeth. "Synthetic Genome Brings New Life to Bacterium," *Science,* Vol. 328, No. 5981, May 21, 2010. http://www.sciencemag.org/content/328/5981/958.full.

Perrett, Bradley. "China Aims for Gas Turbine Catch-up," *Aviation Week,* September 30, 2011. http://www.aviationweek.com/aw/generic/story_generic.jsp?channel=awst&id=news/awst/2011/10/03/AW_10_03_2011_p22–376290.xml&headline=China%20Aims%20For%20Gas%20Turbine%20Catch-up.

Popular Science. *21st Century Soldier.* New York: Time Inc., 2002.

Portree, David S. F. "NASA, Mir Hardware Heritage," March 1995, http://ston.jsc.nasa.gov/collections/TRS/techrep/RP1357.pdf.

Pratt & Whitney Rocketdyne. "RD-180." http://www.pw.utc.com/products/pwr/pro pulsion_solutions/rd-180.asp.

Raibert, Marc, et al. "BigDog, the Rough-Terrain Quaduped Robot." http://www.bos tondynamics.com/img/BigDog_IFAC_Apr-8-2008.pdf.

Rashid, Fahmida Y. "Marine General Calls for Stronger Offense in U.S. Cyber-Security Strategy," *eWeek.com*, July 15, 2011. http://www.eweek.com/c/a/ IT-Infrastructure/Marine-General-Calls-for-Stronger-Offense-in-US-Cyber Security-Strategy-192629/.

Rawnsley, Adam. "Is the Navy Trying to Start the Robot Apocalypse?" *Wired*, March 3, 2011. http://www.wired.com/dangerroom/2011/03/navy-robot-apocalypse/.

Rawnsley, Adam. "MakerBot Commandos: Special Ops Seek 3D Printer," *Wired*, August 12, 2011. http://www.wired.com/dangerroom/2011/08/special-ops-meets-makerbot-commandos-want-3d-printer/.

Reaction Engines LTD. "Current Projects: SKYLON." http://www.reactionengines. co.uk/skylon.html.

Reisinger, Don. "Finland Makes 1Mb Broadband Access a Legal Right," *CNET*, October 14, 2009. http://news.cnet.com/8301–17939_109–10374831–2.html.

Rich, Ben R. and Leo Janos. *Skunk Works*. Toronto: Little, Brown and Company, 1994.

Richelson, Jeffrey T. *Spying on the Bomb*. New York: W.W. Norton & Company, 2006.

Rincon, Paul. "Nanotech Guru Turns Back on 'Goo'," *BBC*, June 9, 2004. http:// newsvote.bbc.co.uk/mpapps/pagetools/print/news.bbc.co.uk/2/hi/science/ nature/3788673.stm.

Rizzo, Albert, et al. "Virtual Reality Goes to War: A Brief Review of the Future of Military Behavioural Healthcare," *Journal of Clinical Psychology in Medical Settings*, 18(2), 176–87.

Rooney, Ben. "The 3D Printer that Prints Itself," *Wall Street Journal*, June 10, 2011. http://blogs.wsj.com/tech-europe/2011/06/10/the-3d-printer-that-prints-itself/.

Rosen, Jacob and Blake Hannaford. "Doc at a Distance," *IEEE Spectrum*, October 2006. http://spectrum.ieee.org/biomedical/devices/doc-at-a-distance.

Roth, Stephen M. "Why Does Lactic Acid Build Up in Muscles? And Why Does it Cause Soreness?" *Scientific American*, January 23, 2006. http://www.scientificamerican. com/article.cfm?id=why-does-lactic-acid-buil.

Sample, Ian. "Pentagon Considers Ear-Blasting Anti-Hijack Gun," *New Scientist*, November 14, 2001. http://www.newscientist.com/article/dn1564.

Sarigul-Klinjn, Marti, et al. "Trade Studies for Air Launching a Small Launch Vehicle from a Cargo Aircraft." American Institute of Aeronautics and Astronautics, 2005, http://www.airlaunchllc.com/AIAA-2005-0621.pdf.

Sauser, Brittany. "Riding an Energy Beam to Space," *Technology Review*, August 5, 2009. http://www.technologyreview.com/blog/deltav/23928/.

Schutte, John, United States Air Force, Human Effectiveness Directorate. "Research-ers Fine-Tune F-35 Pilot-Aircraft Speech System," October 15, 2007, http://www.af.mil/news/story.asp?id=123071861.

Scott, William B. "Two-Stage-to-Orbit 'Blackstar' System Shelved at Groom Lake?" *Aviation Week,* March 6, 2006, www.aviationweek.com.

Shachtman, Noah. "Be More Than You Can Be," *Wired,* Issue 15.03—March 2007. http://www.wired.com/wired/archive/15.03/bemore_pr.html.

Shachtman, Noah. "The Army's New Land Warrior Gear: Why Soldiers Don't Like It," *Popular Mechanics,* October 1 2009. http://www.popularmechanics.com/technology/military/4215715.

Shachtman, Noah. "U.S. Testing Pain Ray in Afghanistan (Updated Again)," *Wired,* June 19, 2010. http://www.wired.com/dangerroom/2010/06/u-s-testing-pain-ray-in-afghanistan/.

Sharklet Technologies, Inc. "Technology." http://www.sharklet.com/technology/.

Shelly, Toby. *Nanotechnology: New Promises, New Dangers.* Nova Scotia: Fernwood Publishing, 2006.

Singer, Peter W. "How to Be All That You Can Be: A Look at the Pentagon's Five Step Plan for Making Iron Man Real," *The Brookings Institution,* May 2, 2008. http://www.brookings.edu/research/articles/2008/05/02-iron-man-singer.

Singer, Peter W. *Wired for War.* New York: The Penguin Press, 2009.

Sofge, Erik. "The Inside Story of the SWORDS Armed Robot "Pullout" in Iraq: Up-date," *Popular Mechanics,* October 1, 2009. http://www.popularmechanics.com/technology/gadgets/4258963.

Sontag, Sherry, Christopher Drew, and Annette Lawrence Drew. *Blind Man's Bluff: The Untold Story of American Submarine Espionage.* New York: Public Affairs, 1998.

Space Exploration Technologies Corporation—SpaceX. "Falcon 9 Launch Vehicle Payload User's Guide." http://www.spacex.com/Falcon9UsersGuide_2009.pdf.

Straziuso, Jason. "US Companies Send Translators To Afghanistan Who Are Old, Out Of Shape, Unprepared For Combat," *Huffington Post,* July 22, 2009. http://www.huffingtonpost.com/2009/07/22/us-companies-send-transla_n_243046.html.

Telemedicine & Advanced Technology Research Center. *Regenerative Medicine: Creating the Future for Military Medicine,* August 2009. http://www.tatrc.org/ports/regenMed/docs/TATRC_regen_med_report.pdf.

Textron Systems. "Sensor Fuzed Weapon." http://www.textrondefense.com/assets/pdfs/datasheets/sfw_datasheet.pdf.

Thompson, Nicholas. "Inside the Apocalyptic Soviet Doomsday Machine," *Wired,* Issue 17.10, September 21, 2009. http://www.wired.com/politics/security/magazine/17-10/mf_deadhand.

Tirpak, John A. "The RPA Boom," *Airforce Magazine,* August 2010. http://www.airforce-magazine.com/MagazineArchive/Documents/2010/August%202010/0810RPA.pdf.

Toffler, Alvin and Heidi Toffler. *War and Anti-War*. New York: Little, Brown and Company, 1993.

Tran, Pierre. "NATO Denies Reported Bomb Shortage in Libya," *Defense News*, April 21, 2011. http://defensenews.com/story.php?i=6295569&c=AIR&s=EUR.

"Trans Atlantic Model." http://tam.plannet21.com/

University of Queensland. "Hyshot." http://www.uq.edu.au/hypersonics/?page=19501.

Van Creveld, Martin. *The Transformation of War*. New York: The Free Press, 1991.

Wallace, Robert and H. Keith Melton. *Spycraft: The Secret History of the CIA's Spytechs, from Communism to Al-Qaeda*. New York: Dutton, 2008.

Walter, Katie. "A New Application for a Weapons Code," *Science & Technology Review*, March 2010. https://str.llnl.gov/Mar10/pdfs/3.10.3.pdf.

Warwick, Graham. "China Targets UAS As Growth Sector," *Aviation Week*, May 5, 2011. http://www.aviationweek.com/aw/generic/story.jsp?id=news/awst/2011/04/25/AW_04_25_2011_p62–312195.xml&channel=defense.

Warwick, Graham. "F/A-18 Shows UCAS-D Can Land On Carrier," *Aviation Week*, July 8, 2011. http://www.aviationweek.com/aw/generic/story_channel.jsp?channel=defense&id=news/asd/2011/07/08/05.xml&headline=F/A-18%20Shows%20UCAS-D%20Can%20Land%20On%20Carrier.

Weinberger, Sharon. "No Pain Ray Weapon for Iraq (Updated and Bumped)," *Wired*, August 30, 2007. http://www.wired.com/dangerroom/2007/08/no-pain-ray-for/.

Weinberger, Sharon. "Terrorist 'pre-crime' detector field tested in United States," *Nature*, 27 May 2011. http://www.nature.com/news/2011/110527/full/news.2011.323.html.

Williams, Mark. "The Total Information Awareness Project Lives On," *Technology Review*, April 26, 2006. http://www.technologyreview.com/communications/16741/.

Wong, Wilson W. S. and James Fergusson. *Military Space Power*. Santa Barbara: Praeger, 2010.

Wright, Austin. "Navy Powers Up Campaign for Great Green Fleet," *Politico*, March 7, 2012. http://www.politico.com/news/stories/0312/73752.html.

Zajac, Andrew. "Robotic arm getting a hand from the FDA," *Los Angeles Times*, February 9, 2011. http://articles.latimes.com/2011/feb/09/nation/la-na-prosthetic-arms-20110209.

Zajtchuk, Russ, Brigadier General, editor, et al. *Medical Aspects of Chemical and Biological Warfare*. Bethesda: Office of the Surgeon General, Department of the Army, United States of America, 1997. http://www.bordeninstitute.army.mil/published_volumes/chemBio/chembio.html.

Index

About the Author

WILSON W.S. WONG is a research fellow with the Centre for Defense and Securities Studies. His first book, *Weapons in Space: Strategic and Policy Implications*, was Volume III in the Centre for Defense and Security Studies' Silver Dart Canadian Aerospace Studies series. He is also the coauthor of *Military Space Power: A Guide to the Issues* with James Fergusson, which was recently published by Praeger/ABC-CLIO. In addition to research support, Mr. Wong contributed illustrations to the Canadian Defense & Foreign Affairs Institute's 2007 *Report on Canada, National Security and Outer Space*.

Emerging military technologies : a guide to the issues / Wilson W. S. Wong.

DATE DUE

PRINTED IN U.S.A.